Radar Engineer's Sourcebook

William Morchin

Artech House
Boston • London

Library of Congress Cataloging-in-Publication Data

Morchin, William C., 1936-
   Radar engineer's sourcebook / William Morchin.

     p.   cm.
   Includes bibliographical references and index.
   ISBN 0-89006-559-4
   1. Radar—Handbooks, manuals, etc. I. Title.

TK6578.M67   1993   92-18686
621.3848—dc20   CIP

© 1993 ARTECH HOUSE, INC.
685 Canton Street
Norwood, MA 02062

617-769-9750

International Standard Book Number: 0-89006-559-4
Library of Congress Catalog Card Number: 92-18686

10 9 8 7 6 5 4 3 2 1

Contents

Preface

This book is an outgrowth of collecting special formulas, graphs, and other data into a personal sourcebook which I often referred to as my "cheat-sheet book." I am sure we all have one like it. The intention is to get a fast answer without all the frills of proof, substantiation, extensive qualification, and story line. The fast answer is the goal of this sourcebook. The user of this sourcebook will find that most of the formulas, graphs, and data are referenced. However, in a few instances, I have lost a reference to a particular item. I apologize for these omissions.

The blank pages at the end of each chapter are here for your use. You can write in your own formulas and data or paste in your own graphs. This sourcebook is designed to be handy and used every day.

The scope of the information in this sourcebook was not intended to be all-encompassing. There are several handbooks published that have complementary or more extensive information such as: *The Radar Handbook*, edited by Merrill Skolnik (McGraw Hill, 1970 [first ed.] and 1990 [second ed.]), *The Spaceborne Radar Handbook*, edited by Leopold Cantafio (Artech House, 1990), *The Microwave Engineers Handbook*, volumes 1 and 2 (Artech House, 1971), *EW Design Engineers' Handbook*, (Horizon House, 1987, 1988, 1990 and 1992), The *Antenna Notebook* by Dick Johnson (Artech House, 1991). Except for the latter two, these handbooks are more text-like than fast-answer type.

The book chapters follow the subject categories of the IEEE radar index system. I have arranged my personal files and notes to follow this index, expanding it as necessary, and suggest it to others. If your article copies and notes are set up with one unified system, you will find a "pigeon hole" to store what otherwise may sit unorganized on your desk or book shelves or in your files.

We hope to expand this sourcebook and the publisher and I welcome suggestions for additions or even contributions of material for a future edition.

WILLIAM C. MORCHIN
Auburn, Washington
January 1992

Chapter 1

Antennas

1.1 General

1.1.1 Effective Antenna Aperture

Effective area is the ratio of received power to the incident power density [3]:

$$A_e = \frac{pq\lambda^2 D(\theta,\phi)}{4\pi} \tag{1.1}$$

where $D(\theta,\phi)$ is the directivity, p is a polarization mismnatch factor, $p = 0.5$ for random polarization, and q is an impedance mismatch factor, $q = 1 - \mid \Gamma \mid^2$. Γ is the input reflection coefficient of the antenna.

1.1.2 Aperture Efficiency

Approximations

Aperture efficiency for linear Taylor array:

$$\eta \approx 1.37 \exp[-(\text{SLL} + 30)/120] \tag{1.2}$$

where SLL is the mainlobe to sidelobe ratio expressed in dB $30 \leq \text{SLL} \leq 60$. The above relationship can be applied to a rectangular aperture by taking the product of the two aperture efficiencies that have been determined separately for the orthogonal illuminations of the aperture.

For general amplitude tapered rectangular and elliptical apertures [6]:

$$\eta \approx 1.1 - 0.013\text{SLL}, \tag{1.3}$$

for a rectangular aperture,

$$\eta \approx 1.23 - 0.013\text{SLL}, \tag{1.4}$$

for an elliptical aperture. Each approximation should be good to at least 60 dB sidelobe levels.

Figure 1.1 shows similar data for the Taylor distribution as well as others [4]-[5]. Taper loss, sometimes called gain factor is defined as the aperture efficiency, expressed in dB. The sidelobe level in the figure represents the level of the highest sidelobe relative to the main beam peak.

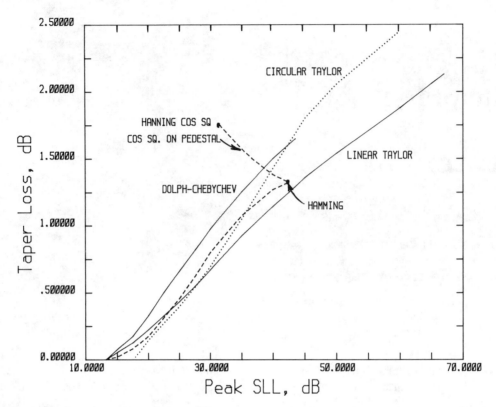

Figure 1.1: Taper loss (aperture efficiency) for a variety of amplitude distributions, after [4] and [5].

Aperture efficiency as determined from planet brightness temperature [7]:

$$\eta = (T_A/T_b)(\lambda^2/\Omega)K \qquad (1.5)$$

where

Ω is the planet solid angle,

T_b is planet brightness temperature,

$$T_A = (\mathrm{T_{obser.}}/\eta_r)\exp(\tau/\sin H)$$

where

$\mathrm{T_{obser.}}$ is the observed antenna temperature,

H is the antenna elevation angle,

τ is atmospheric absorption, where τ is equal to absorption in dB divided by 4.343,

η_r is radome absorption, and

K corrects for planet size relative to antenna beamwidth.

$$K = [1 - (\theta_e/\theta_A)^2]^{1/2}[1 - (\theta_p/\theta_A)^2]^{1/2} \qquad (1.6)$$

where θ_e is the equatorial diameter of the planet, θ_p is the polar diameter of the planet, and θ_A is the antenna half-power beamwidth.

1.1.3 Effective Beamwidth of Combined Transmit and Receive Beams

From [8]:

$$\theta_{\text{eff}} = \sqrt{2}\,\frac{\theta_T\,\theta_R}{\sqrt{\theta_T^2 + \theta_R^2}} \qquad (1.7)$$

where θ_T and θ_R are the half-power beamwidths of the transmit and receive beams, respectively.

1.1.4 Measurement Criteria

Measurement distance effects
Sidelobe change as a function of measurement distance is shown in Figure 1.2.

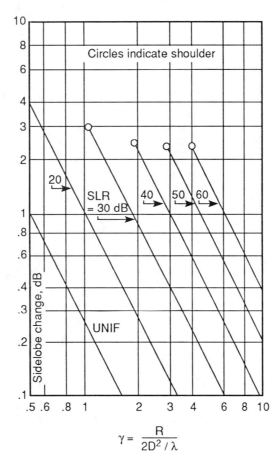

Figure 1.2: Sidelobe change versus normalized measurement distance for Taylor $\bar{n}$ line source, from Hansen [9].

Range distance required [10]

The required measurement distance is shown in Figure 1.3. Range distance required for large antennas [11] taking into account the measurement antenna diameter is given by Uno and Adachi in Figures 1.4, 1.5, and 1.6. R is the measurement distance, D_a is the diameter of the antenna to be measured and D_b is the diameter of the probe or measuring antenna.

For a circular aperture the required distance is [13]:

$$d_{\text{circ}} = \sqrt{\frac{0.2 g_{sl}^{3/8}}{\epsilon_{dB}}} \qquad (1.8)$$

where d_{circ} is the number of $2D_a/\lambda$ distances, g_{sl} is the Taylor numeric mainlobe to sidelobe ratio, a positive value not in dB, and ϵ_{dB} is the sidelobe level change in dB.

1.1.5   Polarization Loss

Response of a receive antenna to a transmit antenna with various combinations of polarization is shown in Table 1.1 [14] below.

Table 1.1: Polarization loss between transmit and receive antennas, from Sefton [14].

RELATIVE dB

TRANSMIT

	V	H	RH	LH	+45°	−45°
V	0	∞	−3	−3	−3	−3
H	∞	0	−3	−3	−3	−3
RH	−3	−3	0	∞	−3	−3
LH	−3	−3	∞	0	−3	−3
+45°	−3	−3	−3	−3	∞	0
−45°	−3	−3	−3	−3	0	∞

(RECEIVE)

Kramer [12] gives a relationship for the loss between an elliptically polarized antenna and a linear polarized antenna as:

$$L_p = 0.5(1 + R\cos 2\tau) \qquad (1.9)$$

where $R = (a_R^2 - 1)/(a_R^2 + 1)$, a_R is the axial ratio of elliptical polarization, and τ is the angle between linear polarization and the major axis of the ellipse. Kramer furthermore predicts the cumulative probability that the polarization loss, L_p, is less than a given value, L, for uniform random over 0 to 2π angular orientation of one antenna relative to the other with:

$$P(L_p \leq L) = (1/\pi)\{\arcsin[(2L-1)/R] + (\pi/2)\} \qquad (1.10)$$

1.1.6   Axial Ratio of Circularly Polarized Antennas with Errors

From Pozar and Targonski [15], the axial ratio due to phase and amplitude errors is shown in Figure 1.7.

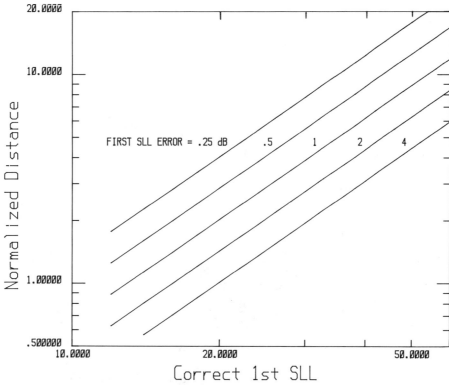

Figure 1.3: Required measurement distance [10].

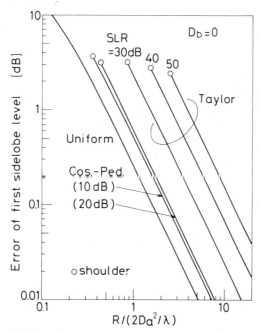

Figure 1.4: Errors of first sidelobe level versus the measurement distance for $D_b = 0$. From Uno and Adachi [11].

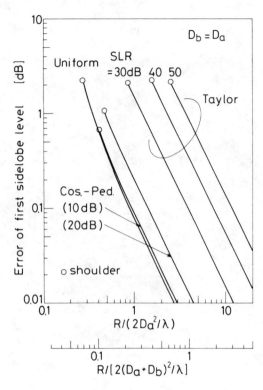

Figure 1.5: Errors of first sidelobe level versus the measurement distance for $D_b = D_a$. From Uno and Adachi [11].

1.1.7   Power Density

Relative power density contours in the $y - z$ plane [16] are shown in Figure 1.8. Aperture diameter is D, z-axis is normal to aperture, y-axis is normal to z-axis. The reader may find these data useful for finding the incident power density on objects in the near field of low sidelobe antennas.

The location of the antenna near field as a function of direction [17] is:

$$R = \frac{(a \cos \theta)^2}{8\Delta} + \frac{a}{2} \sin \theta \qquad (1.11)$$

where R is the outer boundary of the near field, θ is the direction angle from the antenna plane, a is the aperture size, and Δ defines the flatness of the field, typically $\lambda/16$.

1.1.8   Antenna Radar Cross Section

Antenna radar cross section, [19], observed on the antenna axis is:

$$\sigma = G^2 \lambda^2 \Gamma^2 / 4\pi \qquad (1.12)$$

for the radiation mode, not the structural mode. Γ is the reflection coefficient. A grating lobe will be located at:

$$\theta_{RCS} = \arcsin \left[\frac{N}{2S/\lambda} \right] \qquad (1.13)$$

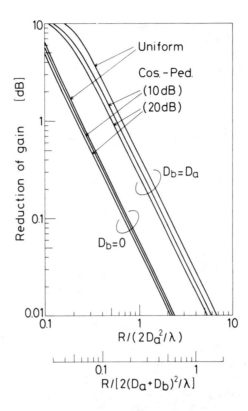

Figure 1.6: Reduction of antenna gain of uniform distribution and cosine-on-a-pedestal distribution versus the measurement distance. From Uno and Adachi [11].

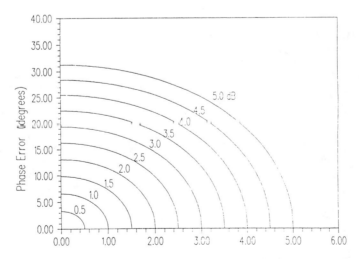

Figure 1.7: Contours of constant axial ratio versus amplitude and phase error. From [15].

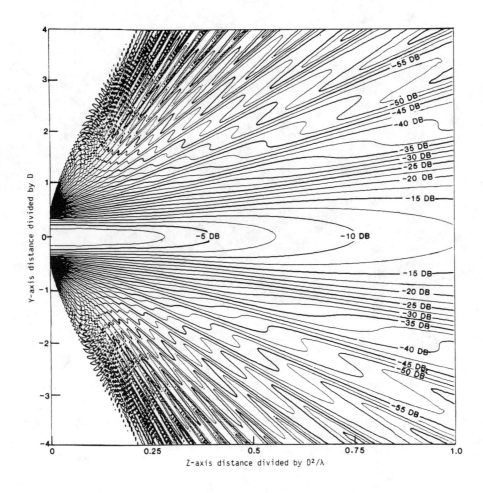

Figure 1.8: Relative power density contours in the y-z plane for a $[1 - (\frac{\ell}{a})^2]^2$ aperture distribution: $D > 30\lambda$, from Lewis and Newell, [16].

N is the grating lobe number, S is the element spacing [19].

1.1.9   Number of Beams in Sphere

The number of mechanically scanned antenna beams in a sphere is:

$$N_b = \pi N_e \tag{1.14}$$

where N_e are number of antenna elements and

$$N_b \approx \frac{\pi}{2} N_e \tag{1.15}$$

for a scanned array.

1.1.10   Approximation to the Fresnel Integral [20]

For the Fresnel integral:

$$F_{\pm}(\chi) = \int_{\chi}^{\infty} \exp(\pm j t^2)\, dt$$

and

$$K_{\pm}(\chi) = \tfrac{1}{\sqrt{\pi}} F_{\pm}(\chi) \exp\left(\mp j \left(\chi^2 + \frac{\pi}{4}\right)\right)$$

$$K_{\pm}(\chi) \approx \left[\frac{1}{2\sqrt{(\pi\chi^2 + \chi + 1)}} + \alpha\chi \exp(-\beta\chi^2) \right]$$

$$\times \exp(\pm j(\arctan(\chi^2 + 1.5\chi + 1) - \tfrac{\pi}{4}))$$

$$\chi \geq 0, \quad \alpha = 0.022, \quad \beta = 0.29.$$

The error in the amplitude component will be within 3.4%.

1.2   Arrays

1.2.1   Rule of Thumb

Edge effects on a low sidelobe antenna can be neglected if array diameter in wavelengths is greater than the required peak sidelobe level [21].

1.2.2   Element Separation and Element Spacing

Element separation, D_s, for a particular scan limit, θ_m in the plane of principle radiation is:

$$D_s \leq \lambda/(1 + |\sin\theta_m|) \tag{1.16}$$

The element areas for rectangular and triangular lattices are related by:

$$\frac{A_t}{A_r} = 0.867 \tag{1.17}$$

1.2.3   Number of Antenna Elements in an Array

For a rectangular grid distribution of elements:

$$N = \frac{A}{\lambda^2 k_s} \tag{1.18}$$

where $k_s = 0.25$ for $\lambda/2$ spacing, $k_s \approx 0.29$ for 0.536λ spacing; the spacing for which a grating lobe will exist at $90°$ from broadside when the array is scanned to $60°$ from broadside.

Also

$$N \approx 0.32G \quad \text{when} \quad k_s = 0.5\lambda \tag{1.19}$$

$$\approx 0.28G \quad \text{when} \quad k_s = 0.536\lambda \tag{1.20}$$

Minimum number of elements:

The minimum number of elements to achieve a gain, $G(\theta_s)$, at scan angle, θ_s, [1] is:

$$N_{\min} \approx 3.02 \times 10^{-4} G(\theta_s)\theta_s^2 \tag{1.21}$$

Scan angle θ_s is expressed in degrees.

1.2.4   Array Gain

Maximum off-boresight array gain

An off-boresight angle, θ_1, exists within the mainbeam such that both an increase and a decrease of the aperture will decrease the gain at that angle. That for rectantular aperture with sides a and b is [17]:

$$G(\theta_1) = \frac{4\pi ab}{\lambda^2} \left\{ \frac{\sin[(\pi a/\lambda)\sin\theta_1]}{(\pi a/\lambda)\sin\theta_1} \right\}^2 \tag{1.22}$$

$G(\theta_1)$ in the plane containing side a is maximum when $\sin\theta_1 = \pm\lambda/2a$ such that

$$G(\theta_1)_{\max} = \frac{4G_o}{\pi^2} \tag{1.23}$$

where G_o is the boresight gain.

Hansen [23] has shown, Figure 1.9, the effect is most pertinent to small arrays. The effect occurs when the scan is in the plane in which the element pattern has directivity. His data pertain to collinear E-plane arrays of half-wave dipoles, with and without a backscreen. We can see that the beam peak angle is always less than the scan angle.

1.2.5   Bandwidth

Bandwidth of microstrip patch antenna element [29]:

$$BW_{\text{MHz}} \approx 128 F_{\text{GHz}} T \tag{1.24}$$

where T is microstrip thickness in inches, F_{GHz} is frequency in GHz, and BW_{MHz} is the bandwidth in MHz for a standing wave ratio ≤ 2.

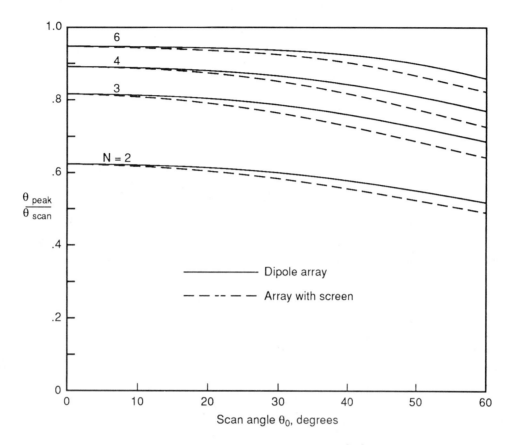

Figure 1.9: Beam peak angle ratio of small scanned arrays, from [23].

1.2.6 Optimum Tilt for Elevation-Scanned Phased Arrays

Figure 1.10 shows a nomograph solution for the determination of the optimum tilt angle of a scanned array.

1.2.7 Design Considerations for Low Sidelobe Microstrip Arrays [24]

Array average pattern with errors:

$$\bar{E}^2(\theta, \phi) = F_e^2(\theta, \phi) \left[E_o^2(\theta, \phi) + \frac{\sigma^2}{D_i} \right] \tag{1.25}$$

where $F_e(\theta, \phi)$ is the directivity of a single element, $E_o(\theta, \phi)$ is the array pattern of isotropic elements, D_i is the directivity of array of isotropic elements, and $\overline{\text{sll}} = D_i/\sigma^2$ is the mainlobe to average sidelobe ratio. Relative to isotropic it is:

$$\text{sll}_i = 1/D_e \sigma^2, \tag{1.26}$$

where D_e is the element directivity.

Sources of error are:

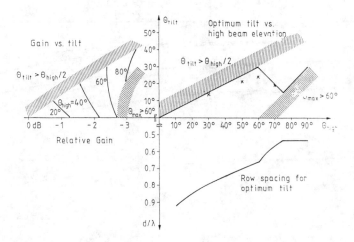

Figure 1.10: Gain at horizon vs. tilt angle (upper left), optimum tilt vs. high beam elevation angle (upper right) row spacing necessary for optimum tilt (lower left). From Solbach [22].

1. Change in resonant frequency of an element:

$$\Delta f_o = \frac{-\Delta L}{L} f_o \qquad (1.27)$$

and

$$= \frac{-\Delta \epsilon_r}{2\epsilon_r} f_o \,; \qquad (1.28)$$

typically 0.5 to 1%.

2. Radiation from feed network:

If printed on the same substrate as the antenna elements the SLL is typically 15 to 30 dB.

3. Mutual coupling.

4. Cross polarization:

Typically caused by positioning errors. To achieve a SLL$_i$ of 20 dB, the cross-polarization of a single element should be -26 dB, assuming an element directivity of 6 dB.

5. Diffraction effects:

For instance, for a 40 dB SLL, a ground plane extending 5λ beyond edge of array is sufficient.

6. Element positioning errors:

$$\begin{aligned} \sigma_\phi &= k_o \Delta R / \sqrt{2} \,, \\ \Delta R &= \sqrt{\Delta X^2 + \Delta Z^2} \,, \end{aligned} \qquad (1.29)$$

where ΔX is the element position error in the aperture plane, and ΔZ is the position error orthogonal to the aperture plane.

7. Imperfect element match and feed network isolation cause amplitude and phase errors with maximum values of:

Maximum amplitude error:

$$\Delta a_{\max} = 1 - \Gamma^2 I^2 . \tag{1.30}$$

Maximum phase error:

$$\Delta \Phi_{\max} = \arctan(\Gamma I) \tag{1.31}$$

where Γ is the magnitude of the reflection coefficient and I is the magnitude of back reflection isolation between feed network ports.

1.2.8   Digital Beamforming

Dynamic range

Dynamic range due to thermal and quantization noise [44]:

$$\mathrm{DR} = \frac{3N 2^{2b}}{26} , \tag{1.32}$$

where

N is the number of array elements,

b is number of digitization bits,

$$\mathrm{DR} = 2^{2(b-1)} N_c$$

from [45], where b is the number of analog-to-digital bits and N_c is the number of parallel channels.

RMS sidelobes

The rms sidelobe level (see also section 1.3.2 subsections containing equations 1.60 through 1.74) due to errors in weight definition (or quantization) [44]:

$$\mathrm{sll}_{rms} = \frac{3N}{2} 2^{2b_w} \tag{1.33}$$

where b_w is the number of weight quantization bits. Due to errors in gain and phase matching [44]:

$$\mathrm{sll}_{rms} = \frac{4N}{\delta^2 + k^2} \tag{1.34}$$

where δ is the phase error in radians either between the in-phase and quadrature channels of any element or the element-to-element mismatch, and k is the ratio of gain error to mean channel gain.

1.2.9   Investment Cost

Cantafio [25] predicts the minimum investment cost for a phased array used for search to be:

$$\text{Minimum cost} = \sqrt{\frac{4(C_t + \bar{P}C_p)C_r\Gamma_s}{\bar{P}}} \tag{1.35}$$

where C_t = cost of building transmit element, \$/element, $\bar{P}$ = Average rf power per element, C_p = cost of producing average rf power, \$/watt, C_r = cost of building receive element, \$/element, and $\Gamma_s = \bar{P}N_T N_R$, where N_T is the number of transmit elements and N_R is the number of receive elements.

From [6], the optimum number of elements for minimum cost for search are:

$$N_{opt} = \sqrt{(C_p \bar{P} A_e k_s)/(C_e \eta_a \lambda^2)}, \tag{1.36}$$

where A_e is the effective aperture area, k_s is the number of elements per square wavelength, η_a is the aperture efficiency, and C_e is the cost of the transmit/receive antenna elements, feed to the elements and ground plane system.

And for track:

$$N_{opt} = (2C_p \bar{P} A_e^2 k_s^2)/(C_e \eta_a^2 \lambda^4)^{1/3} .$$

1.3   Antenna Gain and Patterns

1.3.1   Gain

Gain for a two-dimensional antenna

$$G = \frac{4\pi A_e}{\lambda^2} = \frac{4\pi}{\theta\phi} \tag{1.37}$$

A_e is the effective aperture, θ and ϕ are half-power beamwidths in radians.

Gain for a three-dimensional antenna (e.g., one with end-fire elements)

$$G \approx \frac{4\pi A}{\lambda^2} + \frac{4c}{\lambda} \tag{1.38}$$

where c is the end-fire dimension, taken in the z-direction and the area A is in the orthogonal x- and y-directions [26].

Maximum gain of antenna [27]

$$G_{\max} \approx N^2 + 2N \tag{1.39}$$

where

$$N \approx \frac{2\pi a_{\min}}{\lambda} \tag{1.40}$$

and $a_{\min}$ is the radius of the minimum sized circle to enclose the antenna aperture area.

Gain degradation by element errors

In general:

$$G = \frac{G_o}{1 + \sigma^2} \tag{1.41}$$

where G_o is the gain without errors σ is the rms value of phase and amplitude errors, $\sigma^2 = \epsilon_\phi^2 + \epsilon_A^2/A$ where ϵ_ϕ and ϵ_A are absolute phase, radians, and amplitude errors, respectively.
And:

$$\Delta G = 10 \log \exp \left(\frac{-4\pi\epsilon_{\mathrm{rms}}}{\lambda^2} \right) \tag{1.42}$$

where ΔG is the change in gain, ϵ_{rms} is the rms dimensional change normal to the radiating face.
For a circular aperture [28]:

$$\frac{G}{G_o} = \frac{\sin^2 \varphi + (\cos \varphi - 1)^2}{\varphi^2} \tag{1.43}$$

where φ is the maximum phase error at the aperture edge. For small φ:

$$\frac{G}{G_o} \approx 1 - \pi\delta_\lambda^2 \tag{1.44}$$

where δ_λ is the error in terms of wavelengths.
See also section 1.5.3.

Mainbeam gain function with angle

Using the $\sin x/x$ function:

$$G(x) \approx 20 \log[\frac{\sin(2.783x/\theta)}{2.783x/\theta}] \tag{1.45}$$

Using the Gaussian function:

$$G(x) \approx 20 \log \left[\exp(-1.385x^2/\theta^2) \right] \tag{1.46}$$

Using the cosine function:

$$G(\theta) \approx 20 \log[\cos(90x/\delta)] \tag{1.47}$$

where x is the variable angle, θ is the -3 dB beamwidth, and δ is the beamwidth between the mainlobe and the first null, all in degrees.

Gain of microstrip corporate-fed antennas [29]

$$G \approx 10 \log(4\pi L^2) - \alpha L - C$$

where L is one side of aperture in wavelengths, α is loss, dB, per unit wavelength, and C is a constant due to other mechanisms.

1.3.2   Pattern Calculation Procedure

The sum array factor is [30]:

$$AF_S = (z+1)(z-z_1)(z-z_1^*)\cdots$$
$$(z-z_{M-1})(z-z_M-1^*) \tag{1.48}$$

and the difference array factor is:

$$AF_D = (z-1)\left[\frac{AF_S}{z+1}\right] \tag{1.49}$$

where
$z = \exp(j\Phi)$,
$z_i = \exp(j\Phi_i)$, and $*$ denotes complex conjugate.

$$\Phi = \frac{2\pi d}{\lambda}\cos\theta + \alpha \tag{1.50}$$

α is the progressive phase shift,

d is the element spacing, and

θ is the look angle direction.

$$AF_S = (A_o + A_1 z + \ldots + A_{M-1}z^{M-1})(1 + z^M) \tag{1.51}$$

for $2M$ element array,

$$AF_D = (A_o + A_1 z + \ldots + A_{M-1}z^{M-1})(z^M - 1) \tag{1.52}$$

where A_i are element amplitude coefficients.

The pattern function for a line array is:

$$G(\theta) = 20\log\left\{N(\theta)\sum_n A_n \exp[-j2\pi n(d/\lambda)(s(\theta_p) - s(\theta)]\right\} \tag{1.53}$$

θ is the variate, degrees,
θ_p is the mainbeam pointing angle, degrees, and:

$$N(\theta) = \frac{c(\theta)}{\sum\limits_n A_n}$$

$$c(\theta) = \cos(\theta\pi/180)$$

$$s(\theta) = \sin(\theta\pi/180)$$

A_n is the aperture amplitude for element n.

Sidelobes of rectangular antennas

The absolute gain of sidelobe maxima, e.g. relative to isotropic level, for angles, θ, in the principle plane of radiation beyond the mainbeam is [17]:

For a uniform aperture distribution:

$$G(\theta) = 4/(\pi \sin^2 \theta) \tag{1.54}$$

For a $\cos^2$ aperture distribution:

$$G(\theta) = 4/(\pi \sin^6 \theta) \tag{1.55}$$

For a Hamming aperture distribution:

$$G(\theta) = \frac{0.08 \times 4}{\pi \sin^2 \theta} + \frac{0.92 \times 4}{\pi \sin^6 \theta} \tag{1.56}$$

It is to be noted that the absolute gain of the sidelobes is a function only of direction and is independent of antenna size. The antenna gain is a function of antenna size.

Half-power beamwidth related to sidelobe level (sll)

$$\theta_3 = C \frac{\lambda}{D} \tag{1.57}$$

$C \approx 0.01275 \text{SLL} + 0.73$, from [51],
$\approx 0.0134 \text{SLL} + 0.72$, for a rectangular aperture, from [6],
$\approx 0.0105 \text{SLL} + 0.86$, for an elliptical aperture, from [6].
The approximations from [6] are valid for $20 \leq \text{SLL} < 60$, SLL is expressed in dB.

Sidelobe level for truncated Gaussian distribution

$$\text{SLL} \approx 11.45 - 1.145 E_I \tag{1.58}$$

where SLL is the sidelobe level in dB and E_I is the edge illumination relative to the array center in dB, after data given by [32].

Missing element and 180° phase error effect on array patterns

If the center element of an array is missing, that element pattern adds in phase with the first sidelobe of the array pattern, increasing the combined sidelobe level [51]. For a linear Taylor array the sidelobe level the sidelobe level will be:

$$\text{SLL} = -20 \log(\sqrt{10^{-\text{SLL}_o/10}} + k\sqrt{C/\eta N^2}) \tag{1.59}$$

where SLL_o is the design sidelobe level, $k = 1$, C is the center element power relative to the average power per element (the peaking factor), and η is aperture (taper) efficiency. Values for η, and C are shown in Table 1.2.

If the center element instead has a 180° phase error, as may occur for a failure in a phase shifter, $k = 2$ in the above expression. Figures 1.11 and 1.12 illustrate the effect of an element outage and 180° phase error respectively.

Table 1.2: Taylor Element Peaking Factor and Efficiency, after [51]

Ideal SLL	$\bar{n}$	Efficiency η	Peaking Factor
20 dB	2	0.951	1.664
25	3	.900	1.894
30	4	.850	2.092
35	5	.804	2.262
40	6	.763	2.415
45	8	.726	2.564

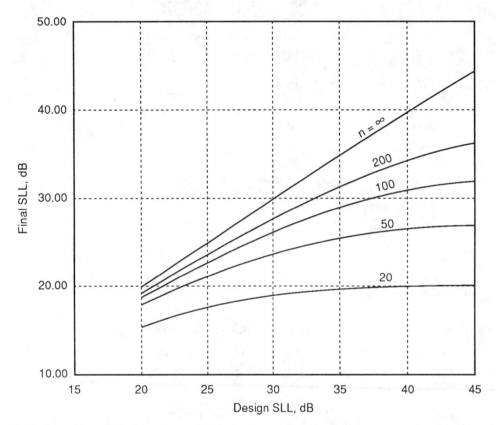

Figure 1.11: Resulting SLL for a linear Taylor array with center element outage, after Howell [51].

Sidelobe dependence on errors

Average amplitude sidelobes:

$$E^2(\theta) = E_o^2(\theta) + \sigma^2/G_o \qquad (1.60)$$

where E is the errored pattern voltage amplitude, E_o is the pattern without error, σ is the rss of the phase and amplitude errors, and G_o is the gain without error.

The rms sidelobe level relative to the mainbeam is:

$$g_{sl} = \frac{\sigma_\Phi^2 + \sigma_A^2}{N_e} \qquad (1.61)$$

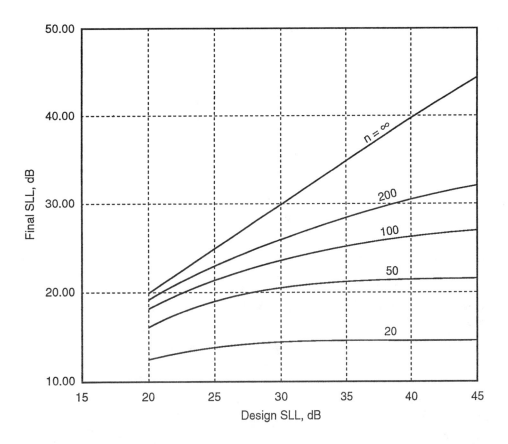

Figure 1.12: Resulting SLL for a linear Taylor array with 180° phase error for center element, after Howell [51].

where σ_Φ is the standard deviation of antenna phase errors, radians:

$$\sigma_\Phi = \sigma_o/57.3$$

σ_A is the standard deviation of antenna amplitude errors.

According to Farrell [34]:

$$g_{sl} = \frac{1}{G_o} - \frac{\theta\varphi}{16\ln 2} \qquad (1.62)$$

where θ and φ are the half-power one-way beamwidths.

The rms sidelobe level dependence on quantized phase shifters:

$$g_{sl} \approx \frac{5}{2^{2P} N_e} \tag{1.63}$$

where P is the number of bits of phase-shift control and N_e is the number of controlled antenna elements.

The binary phase shifter will cause a quantization sidelobe level of [35]:

$$g_{sl} = (2^b - 1)^{-2} \tag{1.64}$$

and a peak phase error of

$$\beta = \pi/2^b$$

the first quantization lobe will be [36]:

$$F_{q1} = \sin \beta/(\pi - \beta) \tag{1.65}$$

and the second lobe will be:

$$F_{q2} = \sin \beta/(2\pi - \beta). \tag{1.66}$$

Sidelobe probability distributions

Normalized array error variance as a function of the ratio of achievable SLL to the designed SLL is shown in Figure 1.13 [37]. The error variance is shown divided by the design SLL. The error variance is $(\epsilon_\Phi)^2 + (\epsilon_A/A)^2$ where ϵ_Φ and ϵ_A are absolute errors in element phase, radians, and amplitude excitation. The abscissa caption, "Deviation from design," is the ratio of achievable SLL to the designed SLL.

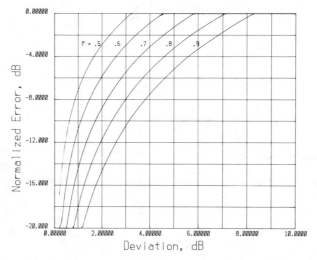

Figure 1.13: Sidelobe level deviation from design for correlated and uncorrelated errors, after Hsiao [37].

As an example, to assure that 90% of the sidelobes do not increase more than 3 dB above the designed sidelobes, the graph shows the error should be no more than about 11 dB relative to the designed sidelobe level (SLL). If the designed sidelobe to mainlobe ratio is -40 dB, the achievable sidelobe to mainlobe ratio will be -37 dB $(-40 + 3)$. The un-normalized variance is -51 dB $(-40 - 11)$, a numeric value of about 8×10^{-5} that cooresponds to an allowable element rms error of about 2.8×10^{-3}. The rms error can be the rss of the correlated and uncorrelated errors, both amplitude and phase.

The curves are applicable to any array and at any point in the sidelobe region.

The curves follow the appoximation:

$$\sigma = 11.26 - 36.7P + (.77 + 10.6P)\ln(\text{SLL}_\delta) \tag{1.67}$$

where P is the cumulative probability of not exceeding an achievable sidelobe level normalized by a design sidelobe level, SLL_δ is the deviation from design, and σ is the array element error normalized by the design sidelobe level.

Sidelobe performance degradation for higher probability values are given by Hsiao [38] in Figure 1.14.

In Figure 1.15 [39], the probability of not exceeding a specified level is plotted against the specified-to-residual ratio for typical choices of design-to-residual ratios. In Figure 1.16, tradeoffs between the residual and design components are plotted for given probability values.

Relationship between rms amplitude and rms phase errors

Lee [40] relates amplitude and phase errors through a parameter, β:

$$\sigma_A^2 = (1 + \beta^2)\exp(-\sigma_\phi^2) - 1 \tag{1.68}$$

where σ_A^2 is fractional amplitude standard deviation, σ_ϕ is the phase standard deviation, radians, and:

$$\beta = \left(\frac{\acute{\sigma}}{\text{sll}_o}\right)\sqrt{2N\eta} \tag{1.69}$$

$\text{sll}_o =$ mainlobe to error-free sidelobe amplitude ratio,
$N =$ number of array elements,
$\eta =$ aperture taper efficiency,
$\acute{\sigma} =$ parameter used in Figure 1.17.

$$\eta = 1/N\Sigma A_n^2, \tag{1.70}$$

A_n are element amplitude weights, shown in Figure 1.17.

The cumulative probability of exceeding the error-free sidelobe level is shown in Figure 1.17 and the relationship between σ_A and σ_ϕ that can be used to determine $\acute{\sigma}$ using equation (1.69) is shown in Figure 1.18.

SLL for systematic errors caused by vibration [41]:

The mainlobe to sidelobe power ratio due to systematic arrors caused by vibration is:

$$\text{SLL}_s \approx (2\pi)^2 \left(\frac{H}{\lambda}\right)^2 \text{SLL}_o \tag{1.71}$$

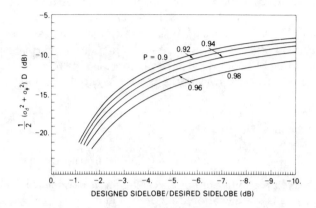

Figure 1.14: Sidelobe degradation for constant probability values, from [38].

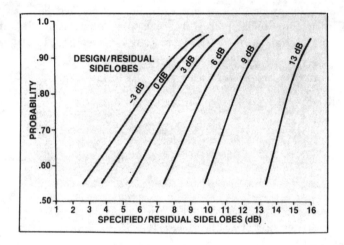

Figure 1.15: Probability function for not exceeding specified sidelobe level, after Kaplan [39].

where H is the average vibration amplitude and SLL_o is the sidelobe level in absence of errors.

For sinusoidal displacement the mainlobe to sidelobe ratio, in dB is [42]:

$$\mathrm{SLL}_{\sin} = -20\log(2\pi a/\lambda) \qquad (1.72)$$

where a is the peak one-way displacement.

The grating lobe location is given by [35]:

$$\sin\theta_{gr} - \sin\theta_{scan} = \lambda/S \qquad (1.73)$$

θ_{gr} is the angle of the grating lobe,

θ_{scan} is the array scan angle,

S is the length of a sinusoidal displacement wave.

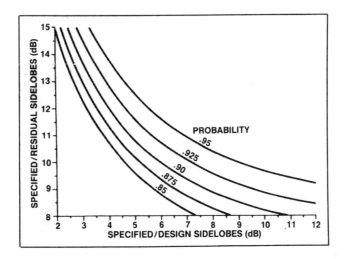

Figure 1.16: Residual and design tradeoff for a given probability, after Kaplan [39].

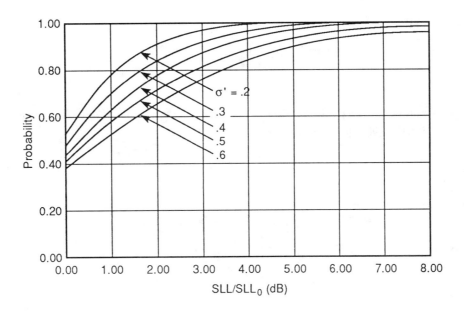

Figure 1.17: Cumulative probability curves for sidelobe degradation estimates, from Lee [40].

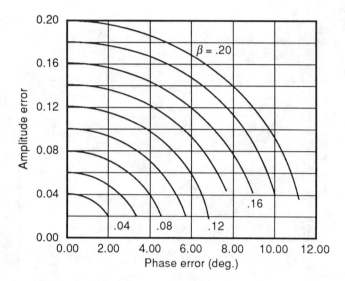

Figure 1.18: Tolerance curves for amplitude and phase errors in an array, from Lee [40].

The beam shift is [43]:

$$\delta = (\lambda/L)(C\Delta\beta/\pi), \text{ radians} \tag{1.74}$$

where L is the array length, $\Delta\beta$ is the peak motion, radians and $C = 1.8$ for a single sinusoid of motion across the array length and $C = 0.26$ for two sinusoids across the array length.

Sidelobe cancellation

An algorithm applicable to a partially adaptive linear array [46] defines the occurrence of sidelobe maxima at:

$$\sin\theta_m = \pm\frac{2m+1}{N+1} \tag{1.75}$$

where $N+1$ is number of elements for even number of elements, $m = 1, 2, \cdots N/2$ is the index of a sidelobe. The sidelobe magnitude at θ_m is:

$$\mid f_m \mid = \left|\sin\left[\frac{\pi}{2}\frac{2m+2}{N+1}\right]\right|^{-1} \tag{1.76}$$

for N even and $\lambda/2$ spacing.

Apply conjugate phase shifts δ and $-\delta$ to a cancellation signal:

$$\delta = -\left[\pi + \frac{\pi}{2}N\frac{2m+1}{N+1}\right] \tag{1.77}$$

with a magnitude:

$$C = \mid f_m \mid /2$$

The required excitation, then, of the edge elements is:

$\mid W_{new}(0^{th} \text{ element}) \mid = \mid W_{new}(N^{th} \text{ element}) \mid$
$\mid W_{new}(N^{th} \text{ element}) \mid = \{[1 - (-1)^m C \cos \delta]^2 + C^2 \sin^2 \delta\}^{1/2}$
and phase:

$$\Phi_N = \arctan\left[\frac{C \sin \delta}{1 - (-1)^m C \cos \delta}\right]$$
$$\Phi_0 = \quad -\Phi_N \tag{1.78}$$

1.4 Radomes

Change of the enclosed antenna beamwidth [6, page 427] can be found from:

$$\frac{\theta_o}{\theta_\epsilon} \approx 1 - \frac{\delta^2}{4\pi} \tag{1.79}$$

where δ is the radome induced quadratic phase error, radians, across the antenna aperture, θ_o is the antenna beamwidth without induced phase error, and θ_ϵ is the beamwidth with induced phase error.

Sidelobe level of the enclosed antenna in the forward sector due to indentations on the radome is:

$$\text{SLL} = -10 \log(\sigma_\phi^2 / G\eta) \tag{1.80}$$

where $\sigma_\phi = 2\pi\tau/\lambda$, and τ is the standard deviation of radome indentations.

Loss of antenna gain is:

$$L_g \approx \exp(-0.6\sigma_\phi)^2 \tag{1.81}$$

Grating sidelobe level due to periodic insertion phase is:

$$\text{SLL}_i = -10 \log(\Delta\phi_r/2), \tag{1.82}$$

dB relative to isotropic where $\Delta\phi_r$ is the periodic phase error.

1.5 Reflector Systems

Komen [33] gives a method for predicting the half-power beamwidth of a parabolic reflector as:

$$\theta = K\lambda/D \quad , \text{ degrees} \tag{1.83}$$

where $K \approx 1.05238 I_{dB} + 55.9486$ and I_{dB} is edge illumination, positive dB.

The reader may want to refer to 1.3.2, eq. 1.58, for prediction of the sidelobe level.

1.5.1 Beam Squint in Parabolic Reflector Antennas

In the direction of propagation, a right-hand-circularly (RHC) feed is radiated left-hand-circularly (LHC) polarized. Similarly a LHC feed is radiated RHC.

For a feed at the focal point and squinted relative to the apex of the reflector the radiated beam will be squinted in a plane that is orthogonal to the feed squint for circular polarized feeds. The squint angle in the orthogonal plane is [47]:

$$\theta_s = \arcsin\left(\frac{\sin \theta_o}{2Fk}\right) \tag{1.84}$$

where θ_o is the feed offset angle, F is the focal length, and k is $2\pi/\lambda$.

For a feed laterally displaced from the focal point:

$$\theta_s = \arcsin\left[\frac{\sin(\theta_F + \theta_B)}{2Fk}\right] \tag{1.85}$$

where θ_F is the squint angle of the feed and θ_B is the angle of the radiated beam in the same plane as the feed squint. If the offset feed is not squinted, $\theta_F = 0$.

1.5.2   Phase Efficiency (Gain Loss) Due to Quadratic Phase Error

For uniform illumination of aperture;

$$\eta_\phi = \frac{\sin^2(\phi) + (\cos\phi - 1)^2}{\phi^2} \tag{1.86}$$

where ϕ is the maximum phase error, in radians, at aperture edge [28].

For ϕ very small $\phi \equiv \delta_r$:

$$\eta_\phi = 1 - \frac{\delta_r^2}{4\pi} = 1 - \pi\delta_\lambda^2 \tag{1.87}$$

where δ_λ is error in terms of wavelength.

$$\delta_\lambda \approx 0.56\sqrt{\frac{\Delta G}{G}} \tag{1.88}$$

$$\eta_\phi \approx [\sin(\phi/2)/(\phi/2)]^2 , \tag{1.89}$$

and from A. Love [48] for a tapered illumination of the aperture:

$$\eta_\phi = \frac{\alpha^2(\cosh\alpha - \cos\phi)}{(\alpha^2 + \phi^2)(\cosh\alpha - 1)} \tag{1.90}$$

where α determines the amount of amplitude taper according to:

$$f(r) = \exp(-\alpha r^2) \tag{1.91}$$

where r is the normalized radial distance from the center of a circular aperture. The edge taper value is given by -8.68α in dB.

Example results are given by Figure 1.19.

1.5.3   Gain Loss Due to Random Surface Deviations

For small values of surface error the loss in antenna gain is:

$$L_s = \frac{G}{G_o} = \exp(-d^2) \tag{1.92}$$

where d is the surface standard deviation error in radians.

$$d = 4\pi\epsilon/\lambda, \tag{1.93}$$

ϵ is the rms surface accuracy.

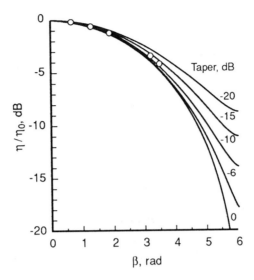

Figure 1.19: Relative loss due to quadratic phase error reaching β rad at aperture edge, from Kramer [48].

For large values of surface error $(d^2 > 10)$ the loss in antenna gain is [12]:

$$L_s = (2c/D_o)/d \tag{1.94}$$

where c is the correlation interval across the aperture and D_o is the aperture diameter.

The difference between large and small surface errors is shown in Figure 1.20.

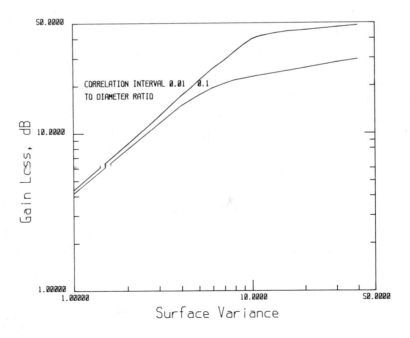

Figure 1.20: Gain loss due to random surface errors in reflector antennas, after [12].

See "Gain degradation by element errors" in 1.3.1 for additional information.

1.5.4   Gain Loss Due to Surface, Jitter, and Pointing Errors

From Safak [50] the antenna gain for pointing and surface errors is:

$$G = G_o \frac{K+1}{K + \exp(4\pi D/\lambda R)^2} \exp[-5.653 \times 10^{-4}(\Delta\theta D/\lambda)^2] \tag{1.95}$$

where

$$K = \frac{1}{\left(\dfrac{4f}{D}\right)^2 \ln\left[1 + \dfrac{1}{(4f/D)^2}\right]} - 1$$

$\Delta\theta$ is the pointing error, $R = D/\epsilon$, D is the parabolic diameter, ϵ is the surface error, and f/D is the parabolic focal length to diameter ratio.

The loss of antenna gain due to jitter is [50]:

$$L_j = (1 + 5.54\delta_{\theta x}^2/\theta_{ox}^2)^{1/2}$$
$$\times(1 + 5.54\delta_{\theta y}^2/\theta_{oy}^2)^{1/2} \tag{1.96}$$

where $\delta_{\theta x}$ and $\delta_{\theta y}$ are the rms angle jitter values in the x and y directions and θ_{ox} and θ_{oy} are the respective half-power beamwidths.

The gain loss for deterministic pointing error is:

$$L_p \approx \exp(-2.77(\Delta\theta/\theta_o)^2) \tag{1.97}$$

where $\Delta\theta$ is the pointing error and θ_o is the half-power beamwidth.

1.5.5   Periodic Error Effects

Sinusoidal radial periodic errors cause sidelobes at discrete angles that increase with the frequency of phase fluctuations caused by the errors. The location of the first such caused sidelobe is given by [42]:

$$\sin\theta = \frac{\lambda}{\pi D}N \tag{1.98}$$

where N is the number of sinusoidal periods in $D/2$.

Azimuthal periodic errors caused by ribs with singularly curved reflectors, called gores, that span the ribs cause sidelobes with locations given by:

$$\sin\theta = \frac{\lambda}{\pi D}1.2N_g \tag{1.99}$$

where N_g is the number of gores.

Azimuthal and radial perodic errors caused by triangular facet reflectors cause sidelobes at locations given by:

$$\sin\theta = \frac{\lambda}{a} \tag{1.100}$$

in each of the three azimuthal planes constraining the facet edges, where a is the facet edge length, and

$$\sin\theta = \frac{\lambda}{a\sqrt{3}} \tag{1.101}$$

in the three planes at 30° to the edges.

Note the approximate relationship between the number of facets, N_f, and the facet size is given in Table 1.3.

Table 1.3: Relationship between number of facets and facet size for triangular facet reflectors.

N_f	a/λ
96	8.6
150	6.9
216	5.8

Bibliography

[1] Jacobs, E., "Design Method Optimizes Scanning Phased Array," Microwaves, April 1982, pp. 69–70.

[2] Yaw, D. F., "Antenna Radar Cross Section," Microwave Journal, September 1984, p. 197.

[3] Francis, M., "Out-of-band response of array antennas," Antenna Meas. Tech. Proc., September 28–October 2, 1987, Seattle, p. 14.

[4] Shrank, H., "Comparison of low sidelobe distributions," IEEE AP Newsletter, Jun 1986, pp. 22–23.

[5] Shrank, H., "Gain factor vs sidelobe level for circular Taylor with optimum n-bar," IEEE AP Newsletter, August 1986.

[6] Morchin, W. C., ed. *Airborne Early Warning Radar*, Artech House, 1990.

[7] Kauffmann, P., *et al.*, "Aperture efficiency of Itapetinga 45-ft. antenna at $\lambda = 3.3$ mm," IEEE Trans. AP-35, No. 4, August 1987, p. 996.

[8] Narayanan, R. M., and McIntosh, R. E., "Millimeter-wave backscatter characteristics of multilayered snow surfaces," IEEE Trans. AP-38, No. 5, May 1990, pp. 693–703.

[9] Hansen, R. C., "Measurement distance effects on low sidelobe patterns," IEEE Trans. AP-32, No. 6, June 1984, p. 593.

[10] Schrank, H., "Measurement of low sidelobe antenna patterns," IEEE AP Newsletter, February 1984.

[11] Uno, T., and Adachi, S., "Range distance requirements for large antenna measurements," IEEE Trans. AP-37, No. 6, June 1989, pp. 707–720.

[12] Kramer, E., "Polarization loss probability," IEEE AP Newsletter, Feb. 1986, pp. 10-11.

[13] Corona, P. *et al.*, "Measurement distance requirements for both symmetrical and antisymmetrical aperture antennas," IEEE Trans. AP-37, No. 8, August 1989, pp. 990–995.

[14] Sefton, H., "Antenna polarization switching," H. Schrank, ed., Antenna Designer's Notebook, IEEE Newsletter, December 1988, p. 17.

[15] Pozar, D. M., and Targonski, S., "Axial ratio of circularly polarized antennas with amplitude and phase errors," IEEE AP Magazine, H. Schrank, ed., Antenna Designer's Notebook, October 1990, pp. 45–46.

[16] Lewis, R., and Newell, A., "An efficient and accurate method for calculating and representing power density in the near-zone of microwave antennas," National Bureau of Standards, NBSIR 85S–3036, 1985.

[17] Cheston, T. C., "Maximum off-boresight antenna gain," IEEE AP Magazine, August 1990, pp. 42–43.

[18] Lalezar, F., and Massey, C. D., "mm-Wave microstrip antennas," Microwave Journal, April 1987, p. 87.

[19] Yaw, D. F., "Antenna radar cross section," Microwave Journal, September 1984, p. 197.

[20] Abuelma'atti, M. T., "An improved approximation to the Fresnel integral," IEEE Trans. AP-37, No. 7, July 1989, pp. 946–947.

[21] Yorinks, L. H., "Edge effects in low sidelobe phased array antennas," IEEE AP Symposium 1985, APS-7-3, p. 225.

[22] Solbach, K., "Optimum tilt for elevation-scanned phased arrays," IEEE AP Magazine, ed. H. Schrank, Antenna Designer's Notebook, April 1990, pp. 39–41.

[23] Hansen, R.C., "Beam peak shift for small scanned arrays," IEEE AP Newsletter, Oct. 1988.

[24] Pozar, D. M., and Kaufman, B., "Design considerations for low sidelobe microstrip arrays," IEEE Trans. AP-38, No. 8, August 1990, pp. 1176–1185.

[25] Cantafio, L. J., "Prediction of the minimum investment cost of phased array radars," IEEE Trans. AES-3, No. 6, November 1967, pp. 207–225.

[26] Huebner, D. A., "Directivity relationships of 3-dimensional antennas," IEEE AP Symposium, 1985, APS-2-1, p. 47.

[27] Harrington, R. F., "On the gain and beamwidth of directional antennas," IRE Trans. AP-6, 1958, pp. 219–225.

[28] Seek, G., "Gain reduction for a circular aperture with quadratic phase error," IEEE AP Newsletter, January 1987, p. 29.

[29] Lalezar, F., and Massey, C. D., "mm-Wave microstrip antennas," Microwave Journal, April 1987, p. 87.

[30] Vu, T. B., "An efficient adaptive phased array design," IEEE National Radar Conference, RADAR 86, Los Angeles, March 12–13, 1986, pp. 35–40.

[31] Howell, J. M., "Phased arrays for microwave landing systems," Microwave Journal, January 1987, pp. 129–137.

[32] Barton, D. K., and Ward, H., *Handbook of Radar Measurements*, Artech House, 1984.

[33] Komen, M., "Microwaves," December 1981.

[34] Farrell, J. L., and Taylor, R. L., "Doppler radar clutter," IEEE Trans. ANE, September 1964, pp. 162–172.

[35] Carpentier, M.H., "Present and future evolution of radars," Microwave Journal, June 1985, pp. 32-63.

[36] Hansen, R. C., "Linear arrays," in: *The Handbook of Antenna Design*, Peter Peregrinus, 1986.

[37] Hsiao, J. K., "On the performance degradation of a low sidelobe phased array due to correlated and uncorrelated errors," International Radar Conference, RADAR-82, London, October 18–20, 1982, pp. 355–359.

[38] Hsiao, J. K., "Optimal design of phased array error tolerance," IEEE International Symposium AP, June 17–21, 1985, APS-4-7, Vol. 1, pp. 127–129.

[39] Kaplan, P.D., "Predicting antenna sidelobe performance," Microwave Journal, Sep. 1986, pp. 201-206.

[40] Lee, J. J., "Sidelobe control of solid-state array antennas," IEEE Trans. AP-36, No. 3, March 1988, pp. 339–344.

[41] Fante,R.L., "Effect of error correlations and systematic errors on average array sidelobes," IEEE Tran AP-38 no.1, June 1990, pp. 124-129.

[42] Corkish, R. P., "A survey of the effects of reflector surface distortions on sidelobe levels," IEEE AP Magazine, December 1990, pp. 6–11.

[43] Nakatsuka, K., "Beam deviations of large linear arrays due to wavy phase errors," IEEE Trans. AP-36, No. 7, July 1988, pp. 1014–1018.

[44] Schoenberger, J.G., Forrest, J. R., and Pell, C., "Active array receiver studies for bistatic/multistatic radar," IEEE International Radar Conference, RADAR 82, London, 1982, pp. 174–178.

[45] Steyskai, H., "Digital beamforming antennas," Microwave Journal, January 1987, p. 107.

[46] El-Azhary, I., *et al.*, "A simple algorithm for sidelobe cancellation in a partially adaptive linear array," IEEE Trans. AP-36, No. 10, October 1988, pp. 1484–1486.

[47] Rahmat-Samii, Y., and Duan, D.-W., "Beam squint in parabolic reflector antennas with circularly polarized feeds," IEEE AP Magazine, February 1990, pp. 31–34

[48] Love, A., "Circular apertures with quadratic phase," H. Schrank, ed., IEEE AP Newsletter, August 1988.

[49] Kramer, E., "Surface accuracy loss," IEEE AP Newsletter, April 1988, pp. 53-54.

[50] Safak, M., "Limitations on reflector antenna gain by random surface errors, pointing errors, and the angle of arrival jitter," IEEE Trans. AP-38, No.1, January 1990, pp. 117–121.

[51] Howell, J.M., "Effects of a missing element on Taylor array patterns," H. Schrank, ed., *Antenna Designer's Notebook*, IEEE AP magazine, December 1990, pp. 32-33.

Notes

Notes

Chapter 2

Array Radar Systems

2.1 Introduction

This chapter summarizes data pertinent to examples of radar engineering accomplishments on array radar systems. Such systems use some form of electronic scanning and normally use a computer to control the antenna beam position. However, electronic scanning may be combined with mechanical.

Electronic scanning is accomplished by controlling the phase of individual array elements or group of elements. The elements may be fed with constrained-fed networks such a corporate systems or fed with an unconstrained system such as a space-fed one.

Active aperture systems that use separate transmit-receive (T/R) modules at each radiating element or each sub array are included in this chapter.

2.2 Transmit-Receive Modules for Array Radar

2.2.1 Transmit-Receive Module Characteristics

From data given by [1], Figures 2.1 through 2.10 show general characteristic trends associated with transmit-receive module development. Some of the transmit-receive module charateristics are summarized with appropriate and simple best-fit equations. As one can see from the plotted data points there can be a substantial deviation about any best-fit equation. In most cases the exponents in the equations are generally accurate, but the leading multiplying values may be changed to offset the best-fit equation either upward or downward to suit the readers purpose. The equations are meant to illustrate only general trends. No equations are suggested for the data in the remaining figures because there are no clear simple trends shown.

From Figure 2.1, we can say:

$$P_{avg} \approx 94 F_{\text{GHz}}^{-2.29}, \qquad \text{W} \tag{2.1}$$

where P_{avg} is the module average rf output power, W, and F_{GHz} is the rf operating frequency, GHz.

And from Figure 2.2:

$$Wt \approx 91.7 P_{avg}^{0.6} \tag{2.2}$$

where Wt is the module weight, g.

Figure 2.5 shows:

$$V \approx 2800 F_{\text{GHz}}^{-2.6}, \qquad \text{cm}^3 \tag{2.3}$$

37

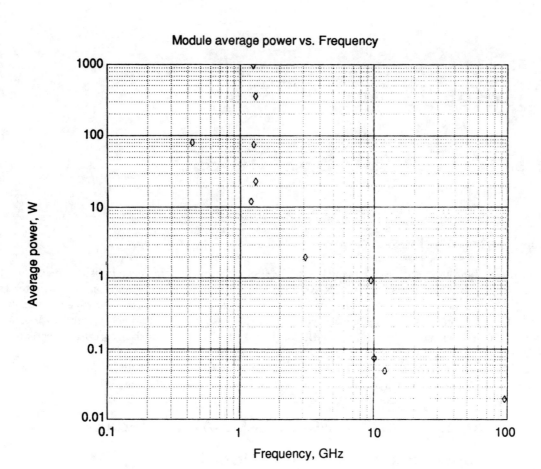

Figure 2.1: Module average power versus frequency.

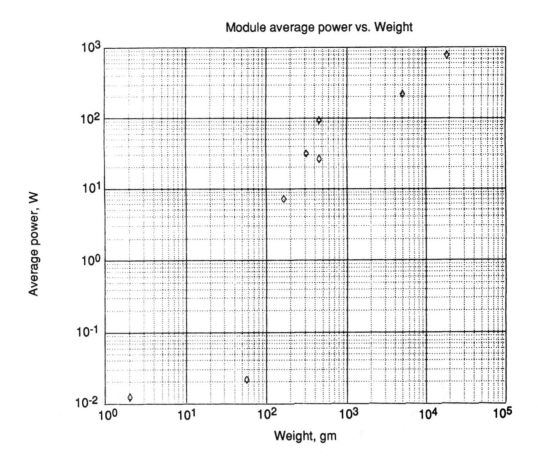

Figure 2.2: Module average power versus weight.

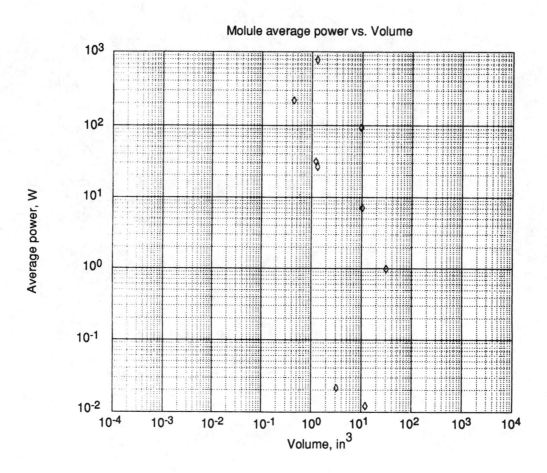

Figure 2.3: Module average power versus volume.

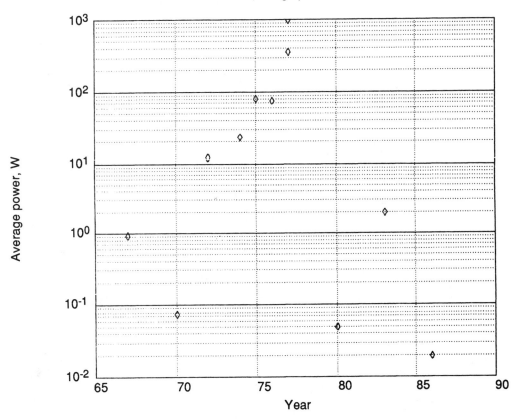

Figure 2.4: Module average power versus year.

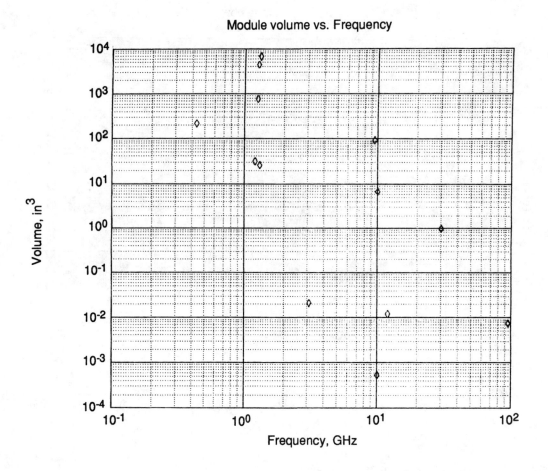

Figure 2.5: Module volume versus frequency.

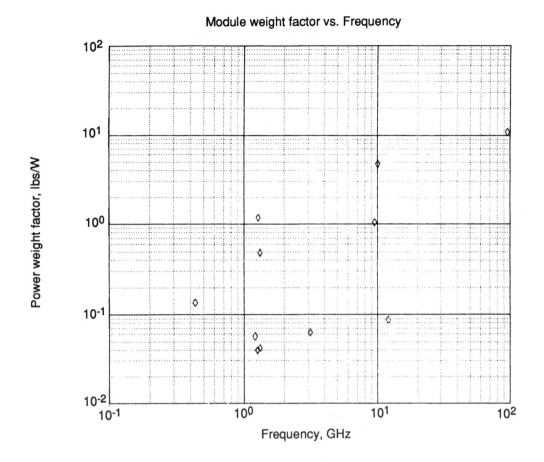

Figure 2.6: Module weight factor versus frequency.

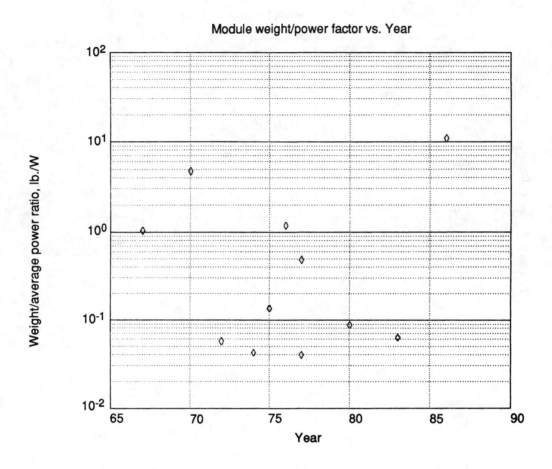

Figure 2.7: Module weight-to-average power versus year.

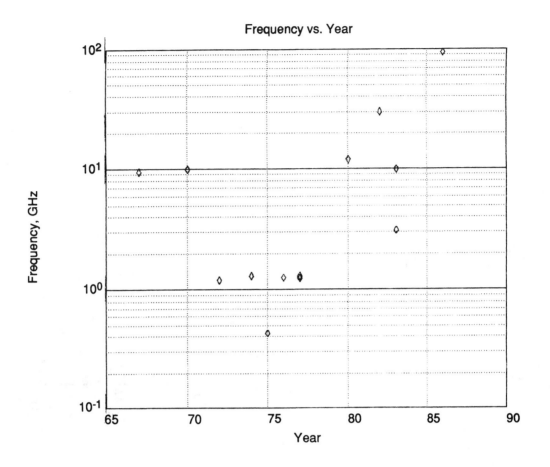

Figure 2.8: Module frequency versus year.

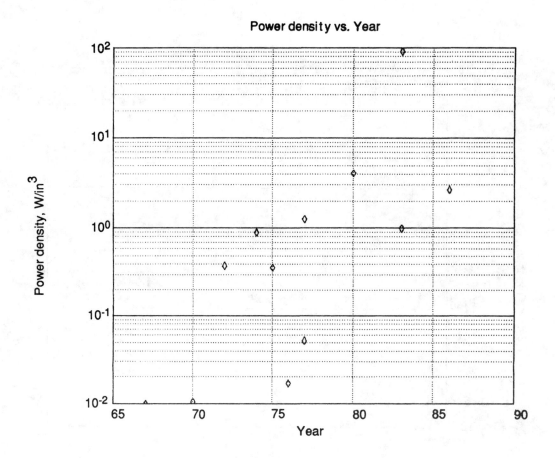

Figure 2.9: Module power density versus year.

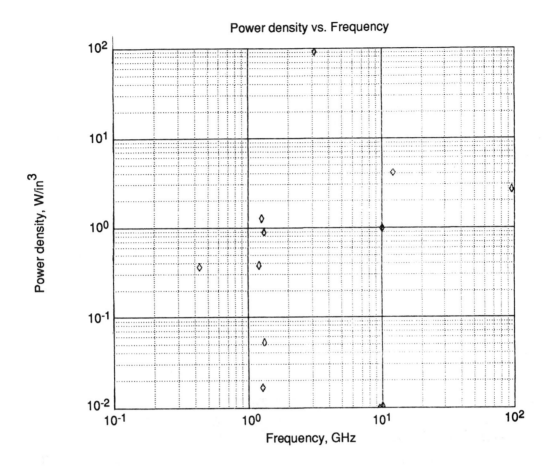

Figure 2.10: Module power density versus frequency.

From Figure 2.6 we see that module weight factor, w_f, g/W is:

$$w_f \approx 51 F_{GHz}^{0.863}, \qquad \text{g/W} \tag{2.4}$$

From Figure 2.9, the change in module power density technology with time is:

$$\rho \approx 4.46 \times 10^{-14} \exp(.3907 Yr), \qquad \text{W/in}^3 \tag{2.5}$$

where Yr is the year, for instance; 89 as in 1989.

T/R Module designs have now progressed to applications of a whole T/R module constructed on a single semiconductor chip. An example set of characteristics of such a T/R module is shown in Table 2.1.

Table 2.1: Example of Ku-band MMIC module performance, [3].

Characterisitic	Value
Center frequency	15 GHz
Tuning range	1.7 GHz
Power output	> 6 mW
Power output at 7 GHz	> 3 mW
Phase noise	< -80 dBc/Hz
(between 100 Hz and 10 MHz)	
DC power dissipation	680 mW
Operation amb.temperature	10 to 40° C
Dimensions	$8 \times 23 \times 3$ mm
Weight	7 g

2.2.2   Example of TR Module Power Consumption

Table 2.2 illustrates the 1986 TR module components that consume power. The inactive and active column headings refer to when the transmit or receive function is being operated. For instance when the T/R module is being used for transmitting the power consumption values in the transmit active and the receive inactive columns are added to determine the power consummed in the transmit mode of operation.

2.2.3   Temperature Distribution of Array [4]

The power absorbed per unit mass is the spatial specific absorption rate:

$$SAR = \frac{\sigma}{2\rho(|E_x|^2 + |E_y|^2 + |E_z|^2)} \tag{2.6}$$

where σ is conductivity of medium,
  ρ is density of medium, and
  E_i is field strength.

Heat resistance in x-direction is:

$$R_x = \frac{1}{k} \frac{\Delta x}{\Delta y \Delta z}, \text{° C/W} \tag{2.7}$$

Table 2.2: Example of the average power consumption of T/R module, W [2].

Technology	Function	Transmit		Receive	
		inactive	active	inactive	active
GaAs hybrid (thin film)	Power Amplifier	2.50	6.75	na	na
Si-hybrid (thin film)	TR power switch	0.03	0.20	0.04	0.12
GaAs monolithic IC	small sig. & LNA	0.25	0.25	0.50	0.50
GaAs and MOS-IC	phase shifter	0.03	0.03	0.07	0.07
GaAs and MOS-IC	micro-attenuator	0.03	0.03	0.07	0.07
GaAs-monolithic IC	small sig. atten.	0.02	0.02	0.04	0.04
MOS & bipolar SI IC	control & monitor	0.05	0.05	0.10	0.10
	Totals	2.91	7.33	0.82	0.90

Total during transmit = 8.15 W
Total during receive = 3.81 W

where Δx, Δy and Δz are the small block dimensions in the array rectilinear coordinates, and k, W/m °C, is thermal conductivity.

Heat capacitance (amount of heat necessary to raise the temperature of mass in block 1°C) is:

$$C = \rho c_p \Delta V = \rho c_p \Delta x \Delta y \Delta z, \text{J/°C} \tag{2.8}$$

c_p is specific heat in J/kg°C, and ΔV is volume of block.

2.2.4 Computer Processing Required

An example is shown in Table 2.3.

Table 2.3: Processing required for typical airborne radar modes.

Airborne Intercept Radar for 1990s	
Signal processing task	Typical computer power Required, MOPS
High PRF search	100
Medium PRF search	250
Doppler beam sharpened map	300
Ground mapping	30
Spotlight map with waypoint detection	200

Bibliography

[1] Lee, C.C., J. Sasonoff, R. Naster, "Transceiver (T/R) module technology for space-based radar," *Space-Based Radar Handbook*, L. Cantafio, ed., Artech House, 1989, Norwood MA.

[2] Langer, E., B. Baulemente, "The mixture of different semiconductor technologies in radar modules and its impact on cooling system problems and reliability," AGARD Conf.Proc. no. 381, Multifunction radar for airborne applications, AGARD-CP-381, AD-A173-978, Jul 1986, pp. 28-1 to 28-8.

[3] Ohira, et al. "Compact full MMIC module for Ku-band phase-locked oscillators," IEEE Tran MTT-37, no.4, April 1989, pp. 723-727.

[4] Zhang, Y., W.T. Jones, J.R. Oleson, "The calculated and measured temperature distribution of a phased interstitual antenna array," IEEE Trans. MTT-38, no.1, Jan 1990, pp. 69-77.

[5] Skolnik, M.I., "Airborne intercept radar for 1990s," Microwave Journal, Jan 1985.

Notes

Notes

Chapter 3

Clutter

3.1 General

3.1.1 Distributions

Conversion of standard deviation values

At times data are gathered or presented in one space, such as log space, whereas it is desired in another space, such as linear. The conversion of mean and standard deviation values between linear and log-normal space is [2]:

For the mean value A:

$$A_{lin} = \exp\left(A_{log} + \frac{\text{SD}_{ln}^2}{2}\right) \tag{3.1}$$

where SD is the standard deviation and subscript *lin* refers to linear space and *ln* refers to log space.

For the standard deviation:

$$\text{SD}_{lin} = \exp(2A_{ln} + \text{SD}_{ln}^2)\exp(\text{SD}_{ln}^2 - 1)$$

Constructing cumulative probability paper

For the Gaussian distribution:

$$x \propto \sqrt{\ln(1-P)^{-1}} - \sqrt{\ln 2}, \text{ for } P > 0.5$$
$$x \propto -\sqrt{\ln P^{-1}} + \sqrt{\ln 2}, \text{ for } P \leq 0.5$$

For the Weibull distribution:

$$x \propto \log[\ln(1-P)^{-1}], \text{ for all } P \tag{3.2}$$

where x defines a position on plotting paper such that functions distributed as above will plot as straight lines for cumulative probabilities, $P < 1$.

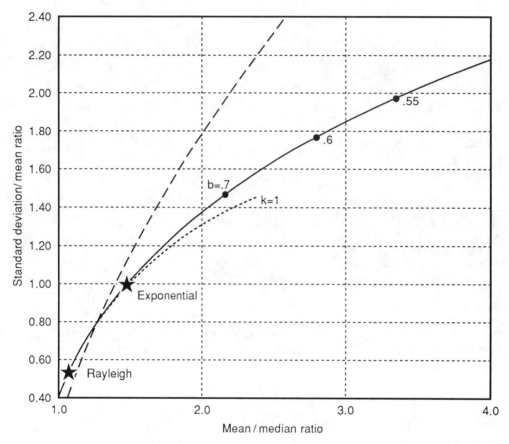

Figure 3.1: Characterization of distributions; --- log-normal,———— Weibull, ---- chi-square, *
Rayleigh and exponential as noted.

Characterizing plots of some distributions

Figure 3.1 shows one way of comparing the characteristics of some common distributions. The ratio
of standard deviation to mean values are compared to the ratio of mean to median values for log-
normal, Weibull, chi-square, exponential and the Rayleigh distributions. The b parameter is shown
for the Weibull distribution. For the Weibull, $b = 1$ corresponds to the exponential, and $b = 2$
corresponds to the Rayleigh. $k = 1$ is shown for the chi-square distribution. For the chi-square,
$k = 2$ corresponds to the exponential. Other values of k can be identified by multiplying the inverse
of the standard deviation to mean ratio by $\sqrt{2}$.

3.1.2   Surface Scattering

Validity of physical optics in rough surface scattering

Papa and Lennon [3] point out that for the traditional geometric optics solution to be valid for
rough surfaces that $R_c \gg \lambda$ and $T \gg \lambda$ when $\sigma/\lambda \gg 1$ (large slope regions) or $\sigma/T < 1$ (small

slope regions). σ is the standard deviation in surface heights, R_c is the average radius of curvature of a surface, and T is the surface correlation length.

$$R_c \approx 0.6T \text{ for a Gaussian surface for } \sigma/T >> 1$$
$$R_c \approx 0.6T/\sqrt{q} \text{ for power law surface where } q \text{ is the exponent.}$$

The Gaussian surface is defined by equations for the correlation coefficient $r(\tau)$:

$$r(\tau) = \sigma^2 \exp(-\tau^2/T^2), \qquad (3.3)$$

and the power-law surface as:

$$r(\tau) = \sigma^2[1 + \tau^2/T^2]^{-q} \qquad (3.4)$$

where τ is the distance between two points on the surface and q is positive.

The average radius of curvature for a Gaussian surface is:

$$R_c = \left(1 + \frac{2\sigma^2}{T^2}\right)^{3/2} \Big/ (12\sigma^2 T^4) \qquad (3.5)$$

$$\text{for } \sigma/T << 1, R_c >> \lambda$$
$$\text{for } \sigma T >> 1, R_c \approx 0.6T$$
$$\text{for } T >> \lambda, R_c >> \lambda$$

Forward scatter

The magnitude of the forward scattered wave [4] is:

$$E_f = \rho_s D R_o E_i \qquad (3.6)$$

where ρ_s is the terrain roughness factor, D is the divergence factor, R_o is the plane earth reflection coefficient, and E_i is the amplitude of the incident wave.

The magnitude of ρ_s is:

$$\rho_s = \exp -\left(\frac{4\pi\Delta h \sin\theta_g}{\sqrt{2}\lambda}\right)^2 \qquad (3.7)$$

where Δh is the rms height of surface irregularities, and θ_g is the grazing angle at the reflection point. This function is shown plotted as curve A in Figure 3.2. There is some concern, [4] that the curve is too steep at the larger grazing angles. Curve B in Figure 3.2 is a modification by multiplying the above equation by, $I_0(\cdot)$, the modified Bessel function of the positive argument of 3.7.

It is to be noted that the effective path difference, in wavelengths, for reflection from a rough surface is:

$$L_\lambda = \frac{2h_{p-p}\sin(\theta_g)}{\lambda} \qquad (3.8)$$

where h_{p-p} is the peak-to-peak height of the surface roughness, and θ_g is the grazing angle.

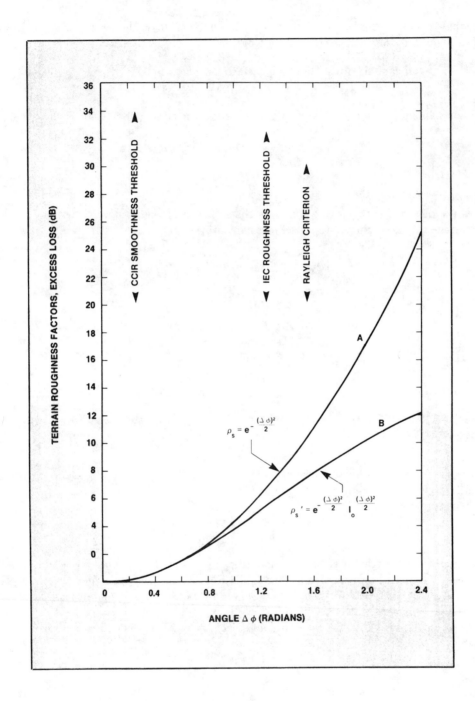

Figure 3.2: Theoretical terrain roughness factor, from [4].

Coherent and incoherent scattering

Eftimiu and Welland have analytically improved the perturbation theory in rough surface scattering studies. Their improvement has extended the range over which the coherent and incoherent components of scattering can be analytically determined. Their results are shown in Figures 3.3 through 3.6. In these figures, the roughness factor; $\xi = (k\sigma)^2 \cos^2 \theta$ where $k = 2\pi/\lambda$, σ is the rms surface roughness, and θ is the angle of incidence to the mean surface. For the figures, $c = kl$, where l is the surface correlation length, $q = \sin\theta_s - \sin\theta$ where θ_s is the scatter angle relative to the normal to the surface. The intensity of the coherent scatter component is one minus the incoherent component shown.

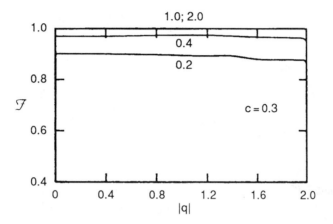

Figure 3.3: Incoherent intensity for horizontal polarization at $c = .3$, and several $k\sigma^2$ values, from [1].

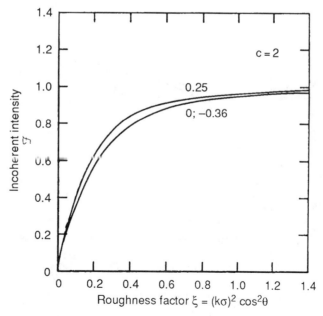

Figure 3.4: Incoherent intensity for horizontal polarization at $c = 2$, several q values, and $\theta = 45°$, from [1].

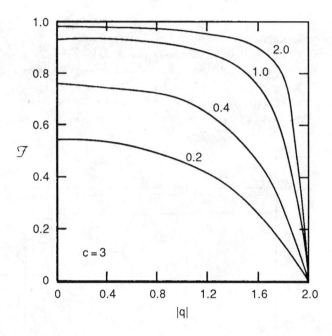

Figure 3.5: Incoherent intensity for horizontal polarization at $c = 3$, and several $k\sigma^2$ values, from [1].

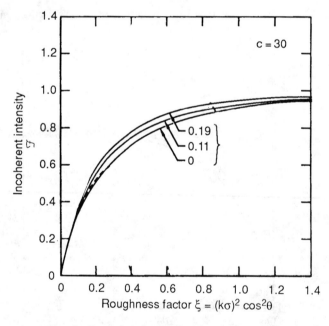

Figure 3.6: Incoherent intensity for horizontal polarization at $c = 30$, several q values, and $\theta = 45^\circ$, from [1].

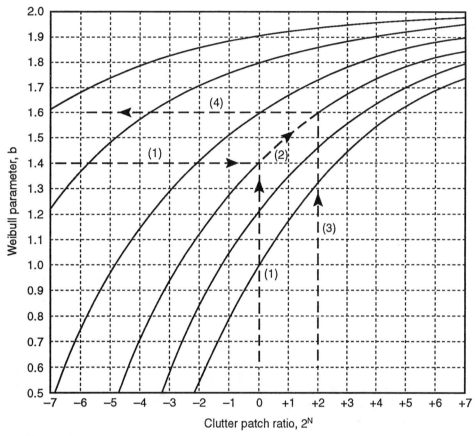

Figure 3.7: Change in Weibull parameter with cell size. Example steps in scaling Weibull parameter are shown in numbered sequence.

3.1.3 Clutter Area for Radar

Pulse length changes

The Clarke and Peters [5] expression for the Weibull parameter change for a doubling of a clutter patch size is:

$$8.5b_{2N} = (7b_N) + 3 \tag{3.9}$$

where b_{2N} is the Weibull shape parameter for a double unit size clutter cell, and b_N is the parameter for a unit size clutter cell. A graphical solution for scaling the Weibull parameter known for one patch size to another is shown in Figure 3.7. The abscissa values, N, can be used to find the scaling ratio from one cell size to another by 2^N. The data have been confirmed in [6] over the range: $-3 \le N \le +3$. For $b = 1$, power will be exponentially distributed and voltage will be Rayleigh distributed. For $b = 2$ the power will be Rayleigh distributed.

To use the graph, find a parametric curve or an extrapolated one for which there is an intersection of the known Weibull parameter and the reference patch size, $N = 0$ on the abscissa. Travel along the parametric curve until reaching a scaled patch size for which the Weibull parameter is desired. The broken lines on the graph show the procedure.

3.1.4   Clutter Spectral Spread

Ulaby [11] presents a formula to estimate the correlation coefficient of clutter for vegetation clutter fluctuation between frequencies of 1 to 18 GHz. His formula is:

$$\rho(\theta, Q) = \exp[-(0.32 + 2.56 \sin\theta)Q] \tag{3.10}$$

where:

$$Q = |\frac{f_2 - f_1}{f_2 + f_1}| \tag{3.11}$$

and θ is the angle of incidence, degrees (0 at the nadir) and f_1 and f_2 are the low and high frequency values of interest.

3.2   Ground Clutter

3.2.1   Land Reflectivity Models

Ulaby and Dobson [7] summarize an extensive body of clutter reflectivity data. They give a function for the mean value reflectivity as:

$$\sigma^o_{\text{mean}}(dB) = P_1 + P_2 \exp(-P_3\theta_i) + P_4 \cos(P_5\theta_i + P_6) \tag{3.12}$$

where θ_i is the angle of incidence, in radians, to the clutter surface and the values of P_i are given in Tables 3.1 through 3.8. Data are provided in the tables with which to estimate the standard deviation of the predicted mean value. They are referred to as $SD(\theta)$ *Parameters* in the tables. The function for calculating the standard deviation is:

$$SD(\theta_i) = M_1 + M_2 \exp(-M_3\theta), \quad dB \tag{3.13}$$

The frequency bands defined by Ulaby [7] for purposes of grouping the clutter data are:

L-Band: 1-2 GHz,
S-Band: 2-4
C-Band: 4-8
X-Band: 8-12
K_u-Band: 12-18
K_a-Band: 30-40
W-Band: 90-100

Ulaby classifies and defines the surface clutter in the tables as:

1. Soil and rock surfaces: barren and sparsely vegetated land. Volcanic material, gravel, desert and non-vegetated surfaces of agricultural land are also included in this category.

2. Trees: forests and orchards. Needle-leaf evergreens, boadleaf trees with leaves, and deciduous broadleaf trees without leaves are included in this category.

Table 3.1: Mean and standard deviation parameter values
for soil and rock surfaces, from [7].

Band	Pol.	Angular Range		δ° Parameters						SD(θ) Parameters		
		θ_{min}	θ_{max}	P_1	P_2	P_3	P_4	P_5	P_6	M_1	M_2	M_3
L	HH	0	50	-85.984	99.0	0.628	8.189	3.414	-3.142	5.600	-5.0×10^{-4}	-9.0
	HV	0	50	-30.200	15.261	3.560	-0.424	0.0	0.0	4.675	-0.521	3.187
	VV	0	50	-94.360	99.0	0.365	-3.398	5.0	-1.739	4.618	0.517	-0.846
S	HH	0	50	-91.20	99.0	0.433	5.063	2.941	-3.142	4.644	2.883	15.0
	HV	0	40	-46.467	31.788	2.189	-17.990	1.340	1.583	4.569	0.022	-6.708
	VV	0	50	-97.016	99.0	0.270	-2.056	5.0	-1.754	14.914	-9.0	-0.285
C	HH	0	50	-24.855	26.351	1.146	0.204	0.0	0.0	14.831	-9.0	-0.305
	HV	0	50	-26.700	15.055	1.816	-0.499	0.0	0.0	4.981	1.422	15.0
	VV	0	50	-24.951	28.742	1.045	-1.681	0.0	0.0	4.361	4.080	15.0
X	HH	0	80	4.337	6.666	-0.107	-29.709	0.863	-1.365	1.404	2.015	-0.727
	HV	0	70	-99.0	96.734	0.304	6.780	-2.506	3.142	3.944	0.064	-2.764
	VV	10	70	-42.553	48.823	0.722	5.808	3.000	-3.142	3.263	11.794	8.977
Ku	HH	0	60	-95.848	94.457	0.144	-2.351	-3.556	2.080	14.099	-9.0	-0.087
	HV	10	50	-99.0	46.475	-0.904	-30.0	2.986	-3.142	5.812	2.0×10^{-4}	-9.0
	VV	0	60	-98.320	99.0	0.129	-0.791	5.0	-3.142	13.901	-9.0	-0.273
Ka	HH											
	HV			Insufficient Data								
	VV											
W	HH											
	HV			Insufficient Data								
	VV											

Table 3.2: Mean and standard deviation parameter loadings for trees, from [7].

Band	Pol.	Angular Range		δ° Parameters						SD(θ) Parameters		
		θ_{min}	θ_{max}	P_1	P_2	P_3	P_4	P_5	P_6	M_1	M_2	M_3
L	HH											
	HV			Insufficient Data								
	VV											
S	HH											
	HV			Insufficient Data								
	VV											
C	HH											
	HV			Insufficient Data								
	VV											
X	HH	0	80	-12.078	1.0×10^{-8}	-10.0	4.574	1.171	0.597	13.111	-9.0	0.073
	HV	0	80	00.000	00.0	-0.636	1.388	6.204	-2.003	12.471	-9.0	-0.125
	VV	0	80	-11.751	2.0×10^{-6}	-10.0	3.596	2.033	0.122	0.816	3.349	0.347
Ku	HH	0	80	-39.042	1.0×10^{-8}	-10.0	30.0	0.412	0.207	13.486	-9.0	-0.154
	HV	0	80	-40.926	1.0×10^{-6}	-10.0	30.0	0.424	0.138	12.614	-9.0	-0.124
	VV	0	80	-39.612	1.0×10^{-6}	-10.0	30.0	0.528	0.023	13.475	-9.0	-0.154
Ka	HH											
	HV			Insufficient Data								
	VV											
W	HH											
	HV			Insufficient Data								
	VV											

Table 3.3: Mean and standard deviation parameter loadings for grasses, from [7].

Band	Pol.	Angular Range θ_{min}	θ_{max}	$\sigma°$ Parameters P_1	P_2	P_3	P_4	P_5	P_6	SD(θ) Parameters M_1	M_2	M_3
L	HH	0	80	-29.235	37.550	2.332	-2.615	5.0	-1.616	-9.0	14.268	-0.003
	HV	0	80	-40.166	26.833	2.029	-1.473	3.738	-1.324	-9.0	13.868	0.070
	VV	0	80	-28.022	36.590	2.530	-1.530	5.0	-1.513	-9.0	14.239	-0.001
S	HH	0	80	-20.361	25.727	2.979	-1.130	5.0	-1.916	3.313	3.076	3.759
	HV	0	80	-29.035	18.055	2.80	-1.556	4.534	-0.464	0.779	3.580	0.317
	VV	0	80	-21.198	26.694	2.828	-0.612	5.0	-2.079	3.139	3.413	3.042
C	HH	0	80	-15.750	17.931	2.369	-1.502	4.592	-3.142	1.706	4.009	1.082
	HV	0	80	-23.109	13.591	1.508	-0.757	4.491	-3.142	-9.0	14.478	0.114
	VV	0	80	-93.606	99.0	0.220	-5.509	-2.964	1.287	2.796	3.173	2.107
X	HH	0	80	-33.288	32.980	0.510	-1.343	4.874	-3.142	2.933	1.866	3.876
	HV	20	70	-48.245	47.246	10.0	-30.0	-0.190	3.142	-9.0	12.529	0.008
	VV	0	80	-22.177	21.891	1.054	-1.916	4.555	-2.866	3.559	1.143	5.710
Ku	HH	0	80	-88.494	99.0	0.246	10.297	-1.360	3.142	2.000	1.916	1.068
	HV	40	70	-22.102	68.807	4.131	-4.570	1.952	0.692	3.453	-2.926	3.489
	VV	0	80	-16.263	16.074	1.873	1.296	5.0	-0.695	-9.0	12.773	0.032
Ka	HH	10	70	-99.0	92.382	0.038	1.169	5.0	-1.906	3.451	-1.118	1.593
	HV											
	VV	10	70	-99.0	91.853	0.038	1.100	5.0	-2.050	2.981	-2.604	5.095
W	HH											
	HV			Insufficient Data								
	VV											

Table 3.4: Mean and standard deviation parameter loadings for shrubs, from [7].

Band	Pol.	Angular Range θ_{min}	θ_{max}	$\sigma°$ Parameters P_1	P_2	P_3	P_4	P_5	P_6	SD(θ) Parameters M_1	M_2	M_3
L	HH	0	80	-26.688	29.454	1.814	0.873	4.135	-3.142	-9.0	14.931	0.092
	HV	0	80	-99.0	99.0	0.086	-21.298	0.0	0.0	4.747	-0.044	-2.826
	VV	0	80	-81.371	99.0	0.567	16.200	-1.948	3.142	-9.0	13.808	0.053
S	HH	0	80	-21.202	21.177	2.058	-0.132	-5.0	-3.142	1.713	3.205	1.729
	HV	0	80	-89.222	44.939	.253	30.0	-0.355	0.526	12.735	-9.0	-0.159
	VV	0	80	-20.566	20.079	1.776	-1.332	5.0	-1.983	2.475	2.308	3.858
C	HH	0	80	-91.950	99.0	0.270	6.980	1.922	-3.142	1.723	3.376	1.975
	HV	0	80	-99.0	91.003	0.156	3.948	2.239	-3.142	13.237	-9.0	-0.178
	VV	0	80	-91.133	99.0	0.294	8.107	2.112	-3.142	1.684	3.422	2.376
X	HH	0	80	-99.0	97.280	0.107	-0.538	5.0	-2.688	2.038	4.238	2.997
	HV	20	70	-28.057	0.0	0.0	13.575	1.0	-0.573	3.301	-0.001	-4.934
	VV	0	80	-99.0	97.682	0.113	-0.779	5.0	-2.076	2.081	4.025	2.997
Ku	HH	0	80	-99.0	98.254	0.098	-0.710	5.0	-2.225	1.941	4.096	2.930
	HV	40	70	-30.403	0.0	0.0	19.378	1.0	-0.590	-9.0	11.516	0.020
	VV	0	80	-99.0	98.741	0.103	-0.579	5.0	-2.210	2.192	3.646	3.320
Ka	HH	20	70	-41.170	27.831	0.076	-8.728	0.869	3.142	2.171	4.391	4.618
	HV											
	VV	20	70	-43.899	41.594	0.215	-0.794	5.0	-1.372	2.117	2.880	4.388
W	HH											
	HV			Insufficient Data								
	VV											

Table 3.5: Mean and standard deviation parameter loadings for short vegetation, from [7].

Band	Pol.	Angular Range		σ° Parameters						SD(θ) Parameters		
		θ_{min}	θ_{max}	P_1	P_2	P_3	P_4	P_5	P_6	M_1	M_2	M_3
L	HH	0	80	-27.265	32.390	2.133	1.438	-3.847	3.142	1.593	4.246	0.063
	HV	0	80	-41.60	22.872	0.689	-1.238	0.0	0.0	0.590	4.864	0.098
	VV	0	80	-24.614	27.398	2.265	-1.080	5.0	-1.999	4.918	0.819	15.0
S	HH	0	80	-20.779	21.867	2.434	0.347	-0.013	-0.393	2.527	3.273	3.001
	HV	0	80	-99.0	85.852	0.179	3.687	2.121	-3.142	13.195	-9.0	-0.148
	VV	0	80	-20.367	21.499	2.151	-1.069	5.0	-1.950	2.963	2.881	4.740
C	HH	0	80	-87.727	99.0	0.322	10.188	-1.747	3.142	2.586	2.946	2.740
	HV	0	80	-99.0	93.293	0.181	5.359	1.948	-3.142	13.717	-9.0	-0.169
	VV	0	80	-88.593	99.0	0.326	9.574	1.969	-3.142	2.287	3.330	2.674
X	HH	0	80	-99.0	97.417	0.114	-0.837	5.0	-2.984	2.490	3.514	3.217
	HV	10	70	-16.716	10.247	10.0	-1.045	5.0	-0.159	-9.0	13.278	0.066
	VV	0	80	-99.0	97.370	0.119	-1.171	5.0	-2.728	2.946	2.834	2.953
Ku	HH	0	80	-99.0	97.863	0.105	-0.893	5.0	-2.657	2.538	2.691	2.364
	HV	0	70	-14.234	3.468	10.0	-1.552	5.0	-0.562	-9.0	13.349	0.090
	VV	0	80	-99.0	97.788	0.105	-1.017	5.0	-3.142	1.628	3.117	0.566
Ka	HH	10	80	-99.0	79.050	0.263	-30.0	0.730	2.059	2.80	3.139	15.0
	HV											
	VV	10	80	-99.0	80.325	0.282	-30.0	0.833	1.970	2.686	-0.002	-2.853
W	HH HV VV	Insufficient Data										

Table 3.6: Mean and standard deviation parameter loadings for road surfaces, from [7].

Band	Pol.	Angular Range		σ° Parameters						SD(θ) Parameters		
		θ_{min}	θ_{max}	P_1	P_2	P_3	P_4	P_5	P_6	M_1	M_2	M_3
L	HH HV VV	Insufficient Data										
S	HH HV VV	Insufficient Data										
C	HH HV VV	Insufficient Data										
X	HH	10	70	-94.472	99.0	0.892	30.0	1.562	-1.918	4.731	-0.007	-3.983
	HV											
	VV	10	70	-59.560	39.284	1.508	30.0	1.104	-1.118	4.260	-0.002	-4.807
Ku	HH	10	70	-90.341	82.900	0.030	1.651	5.0	0.038	5.490	0.001	-6.350
	HV											
	VV	10	70	-38.159	30.320	0.048	1.913	4.356	0.368	6.263	-0.840	0.064
Ka	HH	10	70	-94.900	99.0	0.694	30.0	1.342	-1.718	7.151	-5.201	0.778
	HV											
	VV	10	70	-84.761	99.0	0.797	-30.0	1.597	1.101	3.174	0.001	-0.095
W	HH HV VV	Insufficient Data										

Table 3.7: Mean and standard deviation parameter loadings for dry snow, from [7].

Band	Pol.	Angular Range θ_{min}	θ_{max}	σ° Parameters P_1	P_2	P_3	P_4	P_5	P_6	SD(θ) Parameters M_1	M_2	M_3
L	HH	0	70	-74.019	99.0	1.592	-30.0	1.928	0.905	-9.0	13.672	0.064
	HV	0	70	-91.341	99.0	1.202	30.0	1.790	-2.304	5.377	-0.571	3.695
	VV	0	70	-77.032	99.0	1.415	-30.0	1.720	0.997	4.487	-0.001	-5.725
S	HH	0	70	-47.055	30.164	5.788	30.0	1.188	-0.629	3.572	-2.0×10^{-5}	-9.0
	HV	0	70	-54.390	13.292	10.0	-30.0	-0.715	3.142	13.194	-9.0	-0.110
	VV	0	70	-40.652	18.826	9.211	30.0	0.690	0.214	-9.0	12.516	0.075
C	HH	0	70	-42.864	20.762	10.0	30.0	0.763	-0.147	4.398	0.0	0.0
	HV	0	70	-25.543	16.640	10.0	-2.959	3.116	2.085	13.903	-9.0	-0.085
	VV	0	70	-19.765	19.830	10.0	7.089	1.540	-0.012	13.370	-9.0	-0.016
X	HH	0	75	-13.298	20.048	10.0	4.529	2.927	-1.173	2.653	0.010	-2.457
	HV	20	75	-18.315	99.0	10.0	4.463	3.956	-2.128	2.460	1.0×10^{-5}	-8.314
	VV	0	70	-11.460	17.514	10.0	4.891	3.135	-0.888	12.339	-9.0	-0.072
Ku	HH	0	75	-36.188	15.340	10.0	30.0	0.716	-0.186	3.027	-0.033	0.055
	HV	0	75	-16.794	20.584	3.263	-2.243	5.0	0.096	12.434	-9.0	0.077
	VV	0	70	-10.038	13.975	10.0	-6.197	1.513	3.142	12.541	-9.0	-0.032
Ka	HH	0	75	-84.161	99.0	0.298	8.931	2.702	-3.142	-9.0	13.475	0.058
	HV											
	VV	0	70	-87.531	99.0	0.222	7.389	2.787	-3.142	-9.0	13.748	0.076
W	HH											
	HV											
	VV	0	75	-6.296	5.737	10.0	5.738	-2.356	1.065	3.364	0.0	0.0

Table 3.8: Mean and standard deviation parameter loadings for wet snow, from [7].

Band	Pol.	Angular Range θ_{min}	θ_{max}	σ° Parameters P_1	P_2	P_3	P_4	P_5	P_6	SD(θ) Parameters M_1	M_2	M_3
L	HH	0	70	-73.069	95.221	1.548	30.0	1.795	-2.126	-9.0	14.416	0.109
	HV	0	70	-90.980	99.0	1.129	30.0	1.827	-2.308	4.879	0.349	15.0
	VV	0	70	-75.156	99.0	1.446	30.0	1.793	-2.179	5.230	-0.283	-1.557
S	HH	0	70	-45.772	25.160	5.942	30.0	0.929	-0.284	12.944	-9.0	-0.079
	HV	0	70	-42.940	9.935	15.0	30.0	0.438	0.712	3.276	1.027	8.958
	VV	0	70	-39.328	18.594	8.046	30.0	0.666	0.269	1.157	2.904	0.605
C	HH	0	70	-31.910	17.749	11.854	30.0	0.421	0.740	-9.0	13.0	-0.031
	HV	0	70	-24.622	15.102	15.0	-3.401	2.431	3.142	13.553	-9.0	-0.036
	VV	0	70	4.288	15.642	15.0	30.0	0.535	1.994	4.206	0.015	-2.804
X	HH	0	70	10.020	7.909	15.0	30.0	0.828	2.073	3.506	0.470	15.0
	HV	0	75	4.495	10.451	15.0	-30.0	-0.746	1.083	11.605	-9.0	0.104
	VV	0	70	10.952	6.473	15.0	30.0	0.777	2.081	4.159	0.150	1.291
Ku	HH	0	75	9.715	11.701	15.0	30.0	0.526	2.038	-9.0	13.066	-0.042
	HV	0	75	-79.693	99.0	0.981	30.0	-1.458	2.173	5.631	-1.058	1.844
	VV	0	70	-9.080	13.312	15.0	-4.206	2.403	3.142	-9.0	14.014	0.043
Ka	HH	0	70	43.630	-13.027	-0.860	29.130	1.094	2.802	-8.198	15.0	-0.082
	HV											
	VV	0	70	-33.899	7.851	15.0	30.0	0.780	-0.374	5.488	1.413	0.552
W	HH											
	HV											
	VV	40	75	-22.126	99.0	2.466	0.0	0.0	0.0	4.134	15.0	3.991

3. Grasses: natural rangeland, pasture, hay and small cereal grains such as barley, oats, rye, and wheat. (See following vegetation subsection for additional data.)

4. Shrubs: shrubs, bushy plants, and other crops such as corn, milo, legumes, root crops, alfalfa, canola and cotton.

5. Short vegetation: grass, shrubs, and wetlands. Marshes, swamps, flooded agricultural land, and grass height < 20 cm are included in this category. (See following page.)

6. Road surfaces: man made surfaces such as asphalt, concrete, gravel and dirt.

7. Dry snow: surface layer liquid water content <1% by volume. (See following page.)

8. Wet snow: surface layer liquid water content >1% by volume. (See following page.)

An adaption by Morchin [9] of Barton's [8] land clutter model to account for low incidence angles and the addition of other terrains provides a general land clutter model for UHF through X-band frequencies. The model, corrected and expressed as a numeric value, is:

$$\sigma^o = \frac{A\sigma_c^0 \sin\theta_g}{\lambda} + u\cot^2\beta_0 \exp[-\tan^2\left(\frac{B-\theta_g}{\tan^2\beta_0}\right)]$$

$$u = \sqrt{F_{GHz}}/4.7$$

$$\sigma_c^0 = 1 \, for \theta_g > \theta_c$$

$$\quad = (\frac{\theta_g}{\theta_c})^K \, , \text{ desert only when } \theta_g < \theta_c$$

$$K = 1 \text{ suggested}$$

$$\theta_c = \arcsin(\lambda/4\pi h_e)$$

$$h_e \approx 9.3\beta_0^{2.2}$$

(3.14)

where the constants A, B, β_0, and σ_c^0 are shown in Table 3.9. θ_g is the grazing angle, radians; β_0 represents the rms surface slope; h_e is the rms surface height, and F_{GHz} is the radar frequency, GHz. The constant A can be changed so that the plateau region model reflectivity is modified. B is used to select the general slope of the terrain visible within a clutter cell in the nadir region. It is used to select the central angle of the glistening reflections. The β_0 parameter represents the rms surface roughness. It determines the angular width of the glistening reflections.

Table 3.9: Constants for use in equation 3.13.

Terrain	A	B	β_0	σ_c^0
Desert	0.00126	$\pi/2$	0.14	θ_c/θ_g **
Farmland***	0.004	$\pi/2$	0.2	1
Wooded Hills	0.0126	$\pi/2$	0.4	1
Mountains	0.04	1.24*	0.5	1

* $\pi/2$ less the average mountain slope in a clutter cell should be used. The value suggested here is from [9] for the Cascade mountain range in Washington state of the U.S.

** Use ratio when $\theta_c < \theta_g$, otherwise use 1.

*** See section 3.2.1 for additional data.

Desert

Reference [10] presents a reflectivity model for desert as:

$$\sigma^o = 8.03 \times 10^{-5} \theta_g^{1.5} / \lambda^{0.8} \qquad (3.15)$$

where θ_g is expressed in degrees, σ^o is truncated at lower limit values of 1×10^{-5}, and the leading constant can be changed to fit other terrains.

Vegetation

Ulaby [11] has created an empirical equation to describe the mean and median dependence of clutter reflectivity upon the angle of incidence and frequency for vegetated terrain, in addition to that for short vegetation using equation 3.10. His equation is:

$$\sigma_d^0 = a_0 + a_1 \exp(a_2\theta) + [a_3 + a_4 \exp(-a_5\theta)] \exp[(-a_6 + a_7\theta)F_{GHz}] \qquad (3.16)$$

where the constants are as shown in Table 3.10 and where θ is the angle of incidence, degrees (0 at the nadir), and F_{GHz} is the radar frequency, GHz.

Table 3.10: Parameters for vegetation clutter model, after [11].

Parameters for mean value σ_{dB}^0								
Polarization	a_0	a_1	a_2	a_3	a_4	a_5	a_6	a_7
HH	2.69	-5.35	0.014	-23.4	33.14	0.048	0.053	0.0051
VV	3.49	-5.35	0.014	-14.8	23.69	0.066	0.048	0.0028
HV	3.91	-5.35	0.013	-25.5	14.65	0.098	0.258	0.0021
Parameters for median value σ_{dB}^0								
HH	1.54	-5.36	0.014	-23.8	27.56	0.055	0.152	0.0038
VV	0.699	-4.16	0.016	-17.9	21.15	0.058	0.116	0.0027
HV	-2	-8.89	0.009	-28.9	12.53	0.600	0.256	0.004

Snow surfaces

See equation 3.10 for empirical formulas to determine reflectivity for L- through W-band frequencies.

Narayanan and McIntosh [12] provide algorithms for determining backscatter of a multilayered snow surface at 215 GHz. Figure 3.8 shows dry snow reflectivity and Figure 3.9 shows wet snow reflectivity. The authors conclude that reflectivity is due to both surface and volume mechanisms.

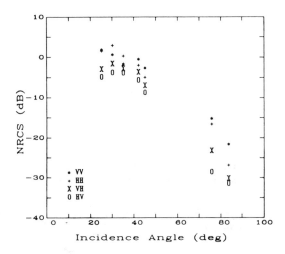

Figure 3.8: Measured values of $\sigma_{VV}^0(*)$, $\sigma_{HH}^0(+)$, and $\sigma_{HV}^0(0)$ of snow at 215 GHz, 0.15% moisture, from [12].

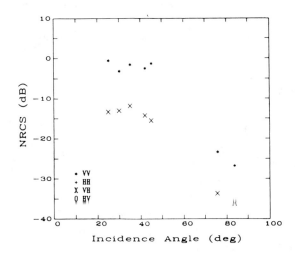

Figure 3.9: Measured values of $\sigma_{VV}^0(*)$, $\sigma_{HH}^0(+)$, and $\sigma_{HV}^0(0)$ of snow at 215 GHz, 1.3% moisture, from [12].

3.2.2 Spectral Spread

From [13] for wooded terrain the rms velocity spread is:

$$\sigma_v \approx .007 W_s^{1.28}, \qquad m/s \tag{3.17}$$

where W_s is the wind speed, m/s.

From [14] for wooded hills:

$$\sigma_v \approx .0045 W_s^{1.4} \tag{3.18}$$

3.2.3 Discretes

Clutter discretes are clutter returns other than diffuse and specular clutter. The source of such returns is, for example, water towers, bridges, buildings, tree trunks, floating debris, and sea-wave prominences [15].

Reference [16] quotes a common model for the relationship between the cross section of a clutter discrete, σ_d (dBm^2), and the number of discretes per mi^2, N_d, that can be stated as:

$$\sigma_d \approx 40 - 10 \log N_d, \text{dB}m^2 \tag{3.19}$$

The leading constant is increased 4 dB if N_d is expressed as the number of discretes per km^2.

Data pertaining to Worcester, Mass. [17] indicate:

$$\begin{aligned}
\sigma_d &\approx 49 - 5 \log N_d, & 49 \leq \sigma \leq 65, \\
&\approx 50 - 13 \log N_d, & 40 \leq \sigma \leq 49
\end{aligned}$$

for grazing angles between 2 and 15 deg. Again the leading constants are increased 4 dB if N_d is expressed in terms of km^2.

3.3 Sea Clutter

3.3.1 Description of Waves and Beaches

Figure 3.10, [18] shows the relationships between the time period of waves, water depth and defines the range of time periods for each wave type.

A sea is the result of superimposing a number of sinusoidal wave trains [18]. The waves in a sea do not have the regular heights of the crests and depths of the troughs. The waves are individual hillocks of water with changing shapes that move nearly independently. There is no particular period, velocity or wavelength. To describe seas statistically, it is customary to use the average value of the highest one-third of the waves, H_3, and the highest one-tenth, H_{10}. The conditions that describe numerically the waves for fully developed seas are shown in Table 3.11. Fetch is the area, taken as length in the table, of open water over which the wind blows to generate waves. The period where most of the wave energy is concentrated is shown in the table is the inverse of the corresponding wave frequency.

Sea state as defined by the World Meteorlogical Organization is shown in Table 3.12.

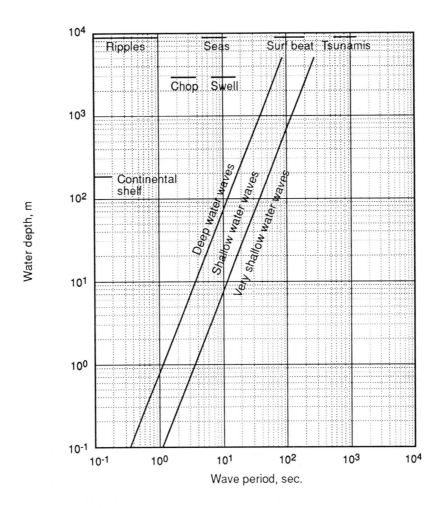

Figure 3.10: Wave period of waves, from [18].

Table 3.11: Conditions in fully developed seas, after [18].

Wind velocity	Fetch	Time	Waves			
			Average height	H_3	H_{10}	Period of most energy
knots	nm	hr	ft	ft		sec
10.0	10.0	2.4	0.9	1.4	1.8	4.0
15.0	34.0	6.0	2.5	3.5	5.0	6.0
20.0	75.0	10.0	5.0	8.0	10.0	8.0
25.0	160.0	16.0	9.0	14.0	18.0	10.0
30.0	280.0	23.0	14.0	22.0	28.0	12.0
40.0	710.0	42.0	28.0	44.0	57.0	16.0
50.0	1420.0	69.0	48.0	78.0	99.0	20.0
m/s	km	hr	m	m	m	sec.
5.1	5.4	2.4	0.3	0.4	0.5	4.0
7.7	18.4	6.0	0.8	1.1	1.5	6.0
10.3	40.5	10.0	1.5	2.4	3.0	8.0
12.9	86.4	16.0	2.7	4.3	5.5	10.0
15.4	151.2	23.0	4.3	6.7	8.5	12.0
20.6	383.4	42.0	8.5	13.4	17.4	16.0
25.7	766.7	69.0	14.6	23.8	30.2	20.0

Table 3.12: Sea description and wind scales, after [18].

Sea State	Wave Height Feet	Meters	Wind Velocity Knots	Descriptive term
0	0	0	0-3	Calm, glassy
1	0-.5	0-.1	4-6	Calm, rippled
2	.5-1.5	.1-.5	7-10	Smooth, wavelets
3	2-4	.6-1.2	11-16	Slight
4	4-8	1.2-2.4	17-21	Moderate
5	8-13	2.4-4.0	22-27	Rough
6	13-20	4.0-6.0	28-47	Very rough
7	20-30	6.0-9.0	48-55	High
8	30-45	9.0-14	56-63	Very high
9	over 45	over 14	≥ 64	Phenomenal

Swells are created when the sea waves move out of a wind area. They are sine wave shapes. The wave velocity is [18]:

$$c = \sqrt{\frac{gL}{2\pi} \tanh \frac{2\pi d}{L}} \qquad (3.20)$$

where g is gravitional constant, L is the wavelength, and d is the depth of water. The wavelength is:

$$L = \frac{g}{2\pi} T^2 \qquad (3.21)$$

where T is the period of the wave.

H/L is wave steepness, where H is the trough to peak wave height. H/L is limited to 1/7 before the wave crests break. Typical characteristic range of values for swells are shown in Table 3.13. The last column also defines the lower limit of deep-water depth above which waves become shallow water waves.

Table 3.13: Approximate lengths and velocities of sinusoidal swell in deep water, after [18].

Period	Wave-length	Velocity	Water depth
(sec.)	(m)	(m/s)	(m)
6.0	56.2	9.4	28.1
8.0	99.9	12.5	49.9
10.0	156.1	15.6	78.0
12.0	224.7	18.7	112.4
14.0	305.9	21.9	152.9
16.0	399.5	25.0	199.8

Beaches

Beaches are classified by slope, [18], as shown in Table 3.14. The bar location shown is the distance from shore where one would expect breakers to occur.

Table 3.14: Beach characteristics [18].

Type	Slope (deg.)	Bar location (m)	Examples
flat	0.5-1	460-1000	Ravenoville, Fr. (Utah Beach N. of WW II) Leadbetter split, WA Katakai, Japan
intermediate	2-6	200-500	Long Branch, NJ, Coos Bay OR
steep	7-10	100-240	Carmel CA, Fort Ord,CA

3.3.2   Sea Reflectivity Models

A general model by Morchin for UHF through X-band frequencies, corrected in format from [9] and expressed as a numeric value, is:

$$\sigma^o = \frac{4 \times 10^{-7} 10^{0.6(SS+1)} \sigma_c^0 \sin\theta_g}{\lambda} + \cot^2\beta_0 \exp\left[\frac{-\tan^2(\pi/2 - \theta_g)}{\tan^2\beta_0}\right]$$

$$\beta_0 = [2.44(SS+1)^{1.08}]/57.29$$

$$\sigma_c^0 = 1 \text{ for } \theta_g > \theta_c$$

$$= \left(\frac{\theta_g}{\theta_c}\right)^K \text{ for } \theta_g < \theta_c \tag{3.22}$$

$$\theta_c = \arcsin(\lambda/4\pi h_e)$$

$$h_e \approx .025 + .046 SS^{1.72} \text{ or}$$

$$\approx H_3/4 \text{ (see tables 3.11 and 3.12.)}$$

$$K = 1 \text{ to } 4 \text{ reported, } K = 1.9 \text{ suggested}$$

where θ_g is the grazing angle, radians; β_0 is the rms surface slope; h_e is the rms surface height.

Abraham [19] presented a set of equations with which to estimate X-band sea reflectivity for either horizontal or vertical polarization. His data are categorized into ranges of grazing angle (in degrees):

For horizontal polarization;

$$\theta_g \leq 1.24 \, \sigma_H^o = 30 + (7 - SS)(8 - SS)/2 - 20\log\theta_g \tag{3.23}$$

$$1.24 < \theta_g \leq 24.33 \; \sigma_H^o = 29 + (7 - SS)(8 - 22)/2 - \theta_g/10$$

$$24.33 < \theta_g \leq 60 \quad \sigma_H^o = (7 - SS)(8 - SS)/2 + 32\cos^2(\theta_g) \tag{3.24}$$

For vertical polarization;

$$\theta_g \leq 1 \qquad \sigma_V^o = 29 + (7 - SS)(8 - SS)/2 - 13\log\theta_g$$

$$1 < \theta_g \leq 7.74 \; \sigma_V^o = 30 + (7 - SS)(8 - SS)/2 - \theta_g$$

$$7.74 < \theta \leq 60 \; \sigma_V^o = (7 - SS)(8 - SS)/2 + 68/3\cos^2\theta_g \tag{3.25}$$

Filippelli, et al. [34] present a more general model, using a reference frequency of 8.9 GHz:

$$\sigma^o = \frac{10^{(m\gamma + b)}}{\lambda^x}$$

$$\gamma \; < 60^o, \text{ but expressed in radians}$$

$$x \; = 0.82 \tag{3.26}$$

$$m \; = 2.4 - .075(k_b - 3) - .42(1 - .0335/\lambda)$$

$$b \; = .2(k_b - 3) - 5.5$$

where k_b is the Beaufort sea state which is 1 more than those shown in Table 3.9 up through sea state 5.

Paulus [20] states that much of the difference between reflectivity models for the low grazing angles can be attributed to evaporation duct effects. This is particularly so for grazing angles less than 1°. Figure 3.11 shows the effect of ducting. The clutter model used in the figure is from Georgia Institute of Technology. It is:

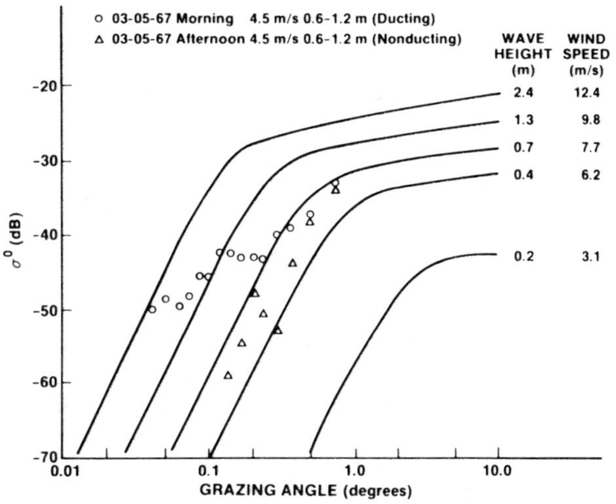

Figure 3.11: Reflectivity measurements made under ducting and nonducting conditions compared to GIT model, from [20].

1) For horizontal polarization:

$$\sigma^o_{HH} = 10\log(3.9 \times 10^{-6}\lambda\theta_g^{0.4}A_iA_uA_w), \ \text{dB} \tag{3.27}$$

A_i is an interference factor;
$A_i = \sigma_\phi^4/(1 + \sigma_\phi^4)$
$\sigma_\phi = (14.4\lambda + 5.5)\theta_g h_{av}/\lambda$
h_{av} is average wave height, m
$h_{av} \approx (V_w/8.67)^{2.5}$

λ is radar wavelength, m
θ_g is grazing angle, radian
V_w is wind speed, m/s

A_u is an upwind/downwind factor;
$A_u = \exp[0.2\cos\phi_w(1 - 2.8\theta_g)(\lambda + 0.02)^{-0.4}]$
ϕ_w is wind direction, deg.

A_w is a wind speed factor
$A_w = [1.9425V_w/(1 + V_w/15)]^{q_w}$
$q_w = 1.1/(\lambda + 0.02)^{0.4}$

2) For vertical polarization, frequency $\geq$ 3 GHz:

$$\sigma^o_{VV} = \sigma^o_{HH} - 1.05\ln(h_{av} + 0.02) + 1.09\ln\lambda + 1.27\ln(\theta_g + 10^{-4}) + 9.7 \tag{3.28}$$

3) For vertical polarization, frequency < 3 GHz:

$$\sigma_{VV}^{o} = \sigma_{HH}^{o} - 1.73\ln(h_{av} + 0.02) + 3.76\ln\lambda + 2.46\ln(\theta_g + 10^{-4}) + 22.2$$

Wind speed determination on the basis of σ^o

Chelton [21] reports on the use of sea clutter measurement of reflectivity to estimate wind speed. The application uses the radar altimeter of the GEOS-3 satellite. The altimeter operates at 13.9 GHz with a 12.5 ns compressed pulse and a 2.6° antenna beamwidth. The relationship used to estimate wind speed at 10 m above the sea surface μ_{10}, m/s, is:

$$\sigma^0 = -2.1 - 10\log(a\ln u_{10} + b), \quad \text{dB} \tag{3.29}$$

$$a = 0.02098 \text{ for } u_{10} < 9.5$$
$$0.08289 \text{ for } u_{10} > 9.8$$
$$b = 0.01075 \text{ for } u_{10} < 9.5$$
$$-0.12664 \text{ for } u_{10} > 9.8$$

3.3.3   Rain Effects on Reflectivity

Hansen [22] presented measurements performed at 9.4 GHz that showed sea reflectivity can increase significantly due to rain splasing on the surface. His results are shown in Figure 3.12. The measurements were obtained using a radar with a 1° beamwidth, a 100 ns pulsewidth at a 2° grazing angle with a surface resolution cell size of 18×30 m.

3.3.4   Sea-Spike Clutter

Any signal rising above a background level can be termed a spike. However, very short isolated periods of high amplitude spikes can be regarded as noise impulses. Olin [23] distinguished sea spikes from background level and noise impulses by defining two time limits and a reflectivity threshold. To qualify as background level the amplitude must persist below a corresponding reflectivity threshold more than 0.1 s. To qualify as a noise impulse the amplitude time period must not exceed 0.01 s. He used a reflectivity threshold of -25 dB for X-band clutter.

Figure 3.13 shows the cumulative probability distribution of sea spike duration. Figure 3.14 shows the cumulative distribution of inter-spike periods. Sea state data and polarization are noted in each figure. An X-band radar with a 1° beamwidth and a 40 ns pulse was used to obtain the data at a grazing angle of 2.9°. The clutter area was 31.6 m^2.

3.3.5   Sea Clutter Spectral Spread

An approximating function for the rms velocity spread, [9] is:

$$\sigma_V \approx 0.101 W_s, \quad \text{m/s} \tag{3.30}$$

where W_s, m/s, is the wind speed for fully deveeloped seas. The corresponding frequency spread is obtained by multiplying by $2/\lambda$.

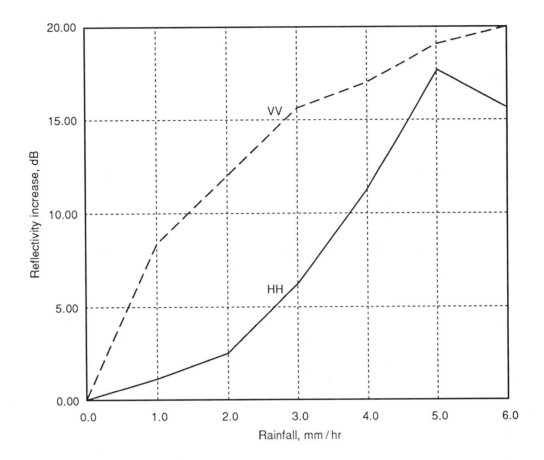

Figure 3.12: Increase in median sea reflectivity due to rain, after [22].

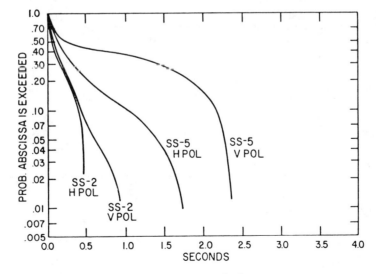

Figure 3.13: Sea spike time duration distribution, from [23].

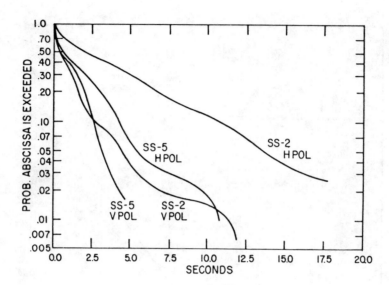

Figure 3.14: Inter-spike time period distributions, from [23].

Sittrop [24] presents an empirical relationship for velocity spread to be expected at X- and Ku-bands:

$$\sigma_v = (\lambda/2)f_o10^{k/10}\,, m/s$$
$$k = \beta\log(2\phi) + [\delta\log(2\phi) + \gamma]\log(W_v/10)$$

where the parameters f_o, β, δ, and γ are given in Table 3.15. W_v is the wind speed, kts and ϕ is incidence angle, deg.

Table 3.15: Experimentally derived constants to determine σ_v, after [24].

Freq. GHz	Wind direction	Polarization	f_o (dB)	β (dB)	δ (dB)	γ
9.4	U	H	60	0	−1.7	7
9.4	C	H	32	2.6	−9.6	15
9.4	U	V	63	0	−7.4	8
9.4	C	V	50	−2.2	−1.6	9.2
16.2	U	H	130	−3	3.6	3.4
16.2	C	H	160	−3	3.3	2
16.2	U	V	125	−4.1	6	4
16.2	C	V	150	−5.4	7	3.4

An example plot is shown in Figure 3.15.

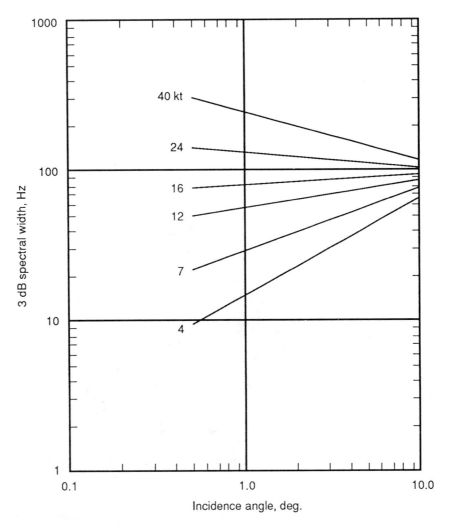

Figure 3.15: Sea clutter velocity spread and spectrum bandwidth at X-band for crosswind and horizontal polarization, from [24].

Hf sea clutter spectral spread

An example Doppler spectrum for a 9.25 MHz HF ground-wave radar is shown in Figure 3.16. The two peaks are the approaching and receding first order Bragg lines. Bragg scattering is reinforcing scattering from periodic waves.

Bragg scattering occurs when the sea wavelength, L matches the radar wavelength:

$$L = n\lambda/2\sin\theta_i\,, n = 1, 2, \ldots \tag{3.31}$$

where θ_i is the incidence angle

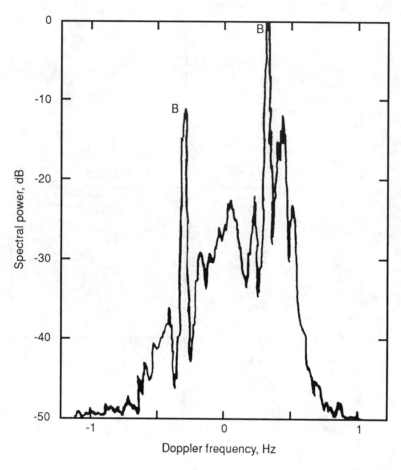

Figure 3.16: Doppler sea-echo spectrum obtained with 9.25 MHz ground-wave radar, after [26].

The Doppler frequency is [25]:

$$f_s = \pm\sqrt{ngf_o\sin\theta_i/\pi c} \qquad (3.32)$$

where g is acceleration of gravity, f_o if the radar frequency, c is light velocity.

3.4   Volume Clutter

3.4.1   Rain Backscatter

Rain reflectivity in terms of m^2 per m^3 of rain volume is:

$$\eta = \frac{\pi^5}{\lambda^4}K^2Z^{-18} \qquad (3.33)$$

K^2 is a factor involving the complex refractive index of water. It varies with temperature and frequency. Some typical values are shown in Table 3.16.

The reflectivity factor Z varies with rain rate, and rain attenuation. A nominal value is: $Z = 296R^{1.47}$, where R is the rain rate, mm/hr. Other specific values are given in Table 3.17.

Table 3.16: Factor K dependence on temperature and frequency, after [28].

Temperature (°C)	Frequency (GHz)			
	3	9.3	24	48
20	.928	.9275	.9193	.8926
0	.934	.9300	.9055	.8312

Table 3.17: Relation of reflectivity factor, Z, and rain rate, R, after [27].

Frequency (GHz)	Rain Interval (mm/hr)	Reflectivity Factor Z
3		$295R^{1.45}$
5.5		$280R^{1.45}$
6.4		$280R^{1.45}$
9.3		$275R^{1.55}$
16	0–20	$330R^{1.54}$
	20–50	$500R^{1.5}$
	50–100	$750R^{1.3}$
24	0–5	$356R^{1.5}$
	5–20	$460R^{1.35}$
	20–100	$820R^{1.15}$
35	0–5	$350R^{1.32}$
	5–20	$450R^{1.15}$
	20–100	$780R^{0.95}$
48	0–5	$240R^{1.1}$
	5–20	$345R^{0.9}$
	20–100	$540R^{0.75}$

3.4.2 Aurora Clutter Characteristics

The aurora zone is shown in Figure 3.17 [32]. The magnetic dip angle is the angle of the earth's magnetic lines relative to a plane normal to the magnetic axis. Occurance of aurora is periodic with each day and is most frequent in morning and evening hours.

Volume spectral reflectivity of aurora is approximated in Figure 3.18 from data presented in several sources, [29], [30], [31]. The data are sensitive to the angle of incidence to the earths magnetic lines. The reflectivity decreases and become diffuse, from that shown, about 10 dB/degree of aspect change from the perpendicular to the magnetic lines. To encounter specular aurora clutter one must be a large distance from the auroral oval.

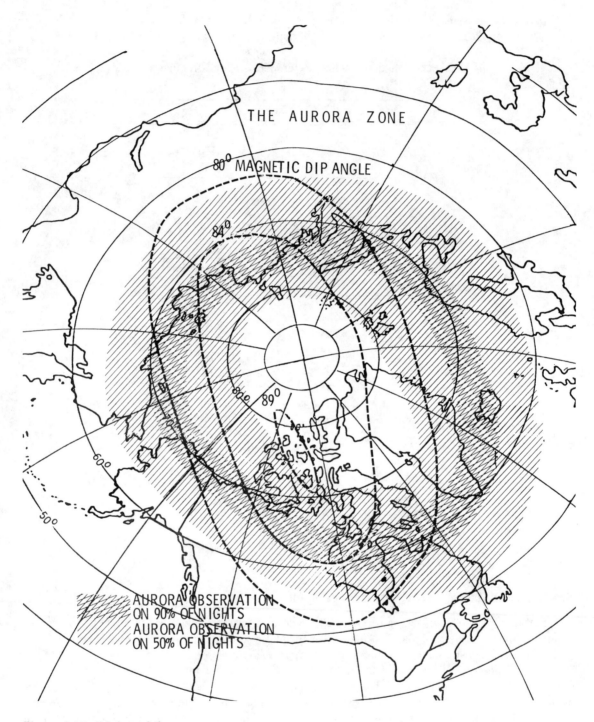

Figure 3.17: Region of the aurora zone.

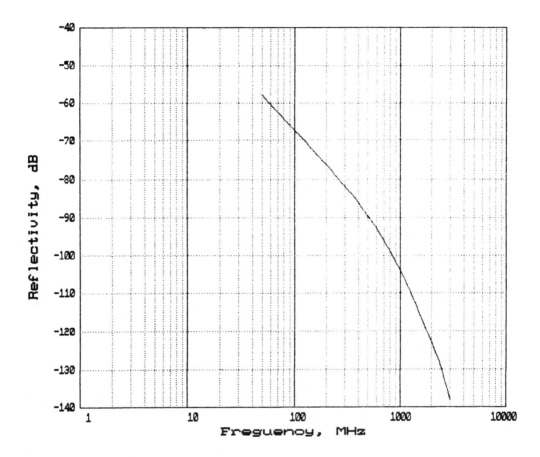

Figure 3.18: Aurora volume reflectivity.

Auroral clutter spectral characteristics

Figure 3.19 shows by example the great uncertainity of the aurora diffuse clutter spectral characteristics, [33] that can be expected. The three curves represent data that were measured in sequence a, b, and c within a time span of 3-seconds in a volume 20 × 20 × 100 km. The Doppler velocities shown were computed for a 398 MHz radar located at Homer, Alaska.

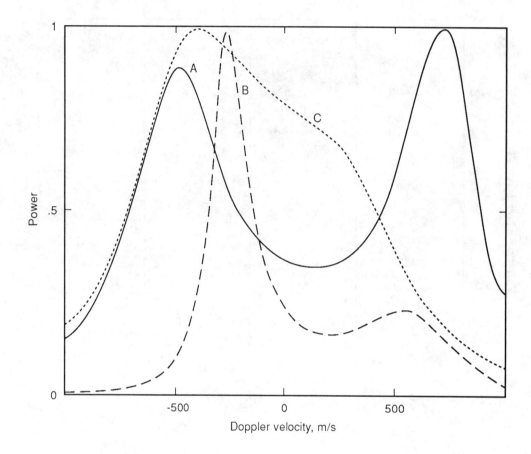

Figure 3.19: Example spectral characteristics of diffuse aurora, after Moorcroft [33].

Bibliography

[1] Eftimiu, C., C. G. Welland "The use of Pad' approximations in rough surface scattering," IEEE Tran. AP-35, no. 6, June 1987, p. 721.

[2] Lane, T.L., et al. "An overview of MMW radar reflectivity," Proc. IEEE National Radar Conf., Apr. 20-21, 1988, p. 239.

[3] Papa, R.J., J.F. Lennon, "Conditions for the validity of physical optics in rough surface scattering," IEEE AP-36, no.5, May 1988, pp. 647-650.

[4] Vogel, W.J., E.K. Smith, "Propagation considerations in land mobile satellite transmission," Microwave Journal, Oct. 1985, pp. 111-130.

[5] Clarke, J., R.S. Peters, "The effect of pulse length changes on Weibull clutter," Royal Radar Establishment, Memorandum No. 3033, Nov. 1976.

[6] Chen, P.W., W.C. Morchin, "Detection performance with multiple background distribution," IEEE NAECON Conf., Dayton, OH, May 1978, pp. 1309-1311.

[7] Ulaby, F.T., M.C. Dobson, *Handbook of Radar Scattering Statistics for Terrain*, Artech House, Norwood, MA, 1989.

[8] Barton, D.K., "Land clutter models for radar design and analysis," Proc. IEEE, vol. 73, no. 3, pp. 198-204.

[9] Morchin, W.C., *Airborne Early Warning Radar*, Artech House, Norwood, MA, 1990, p. 150.

[10] Hayes, R.D., *White paper 295-34890*, Georgia Institute of Technology, 1981.

[11] Ulaby, F.T., "Vegetation clutter model," IEEE Tran AP-28, no.4, Jul. 1980, pp. 538-545.

[12] Narayanan, R.M., R.E. McIntosh, " Millimeter-wave backscatter characteristics of multilayered snow surfaces," IEEE Tran AP-38 no.5, May 1990, pp. 693-703.

[13] Nathanson, F.E., *Radar Design Principles*, McGraw Hill, New York, 1969, p.272.

[14] Barton, D.K., *Radar Systems Analysis*, Artech House, Norwood, MA, 1977, p.100.

[15] Long, M.W., *Radar Reflectivity of Land and Sea*, Artech House, Norwood, MA, 1983, p.xviii.

[16] Long, W.H., D.H. Mooney, W.A. Skillman, "Pulse doppler radar," Ch. 17 of *Radar Handbook*, 2nd ed., M.I. Skolnik, ed., McGraw Hill, 1990, p.17-11.

[17] "Data sheet," Mitre Corp., circa 1972.

[18] Bascom, W., *Waves and Beaches*, Doubleday and Co., Garden City, NY, 1964.

[19] Abraham, D., "Numerically modeling sea backscatter coefficients," *Microwaves and RF*, Apr. 1983.

[20] Paulus, R.A., "Evaporation duct effects on sea clutter," IEEE Trans. AP-38, no.11, Nov. 1990, pp. 1765-1771.

[21] Chelton, D.B., "Comment on seasonal variation in windspeed and sea state form global satellite measurements," Jour. Geo. Phys. Research, 1985 Vol. 90, May 1985.

[22] Hansen, J.P., "Rain backscatter tests dispel old theories," Microwaves & RF, June 1986, pp. 97-102.

[23] Olin, I.D. "Amplitude and temporal statistics of sea spike clutter," IEE Inter. Conf. RADAR-82, London, Oct.18-20, 1982, pp. 198-202.

[24] Sittrop, H., "On the sea-clutter dependency on windspeed," Int. Conf. RADAR-77, London, Oct. 1977, pp. 110-114.

[25] Kingsley, S.P., "The coherence of HF radar sea echoes," GEC Jour. Res., vol. 4, no. 3, 1986, pp. 203-210.

[26] Clarke, J., D.E.N. Davies, M.F. Radford, "Review of United Kingdom radar," IEEE Trans. AES-20, no. 5, Sep. 1984, pp. 506-520.

[27] Bogush, A.J., *Radar and the Atmosphere*, Artech House, Norwood, MA, 1989.

[28] Tank, W.G., "Atmospheric effects," Ch.3, ed. W.C. Morchin, *Airborne Early Warning Radar*, Artech House, Norwood, MA, 1990.

[29] Hodges, J., R.L. Leadabrand, "Auroral radar echo wavelength dependence," Jour. Geophysical Res., vol. 71 no. 19, Oct. 1966, pp. 4545-4549.

[30] Minkoff, J., "Calculation of per-unit- volume rf scattering cross-sections in radar aurora," Canadian Jour. of Physics, vol. 56, no. 2, 1978, pp. 280-287.

[31] Chesnut, W.G., J.C. Hodges, R.L. Leadabrand, "Auroral radar backscatter study" IEEE Int. Union Radio Science Spring Meeting, Washington, DC, Apr. 21-24, 1969.

[32] Valley, S.L. (ed.) *Handbook of Geophysics and Space Environments*, McGraw-Hill, New York, 1965.

[33] Moorcroft, D.R., "Rapid scan doppler velocity maps of the UHF diffuse radar aurora" Jour. Geo. Res., vol. 83, no. A4, Apr. 1978, pp. 1482-1492.

[34] Filippelli, L.J., et al. "A generic space spaced based radar model with resource scheduler," AD-A126639, NRL no. 5059, Apr. 1983.

Notes

Notes

Chapter 4

Cross Section

4.1 Definition and Statistics

4.1.1 Definition

From [5] we can express the radar cross section as:

$$\sigma = 4\pi \frac{P_s}{P_i} \tag{4.1}$$

where P_s is the per unit solid angle scattered in a specified direction. P_i is the power per unit area in a plane wave incident on the scatterer from a specified direction. If not otherwise stated, it can be assumed that the transmitting and receiving antennas have the same polarization, the entire target is uniformly illuminated, the pulse is long enough to completely cover the target, and the receiving antenna is in the far-field of the target.

Scattering at relativistic velocities

From the special theory of relativity, a wave reflected from a uniformily moving mirror has a reflected angle, θ, [7]:

$$\cos\theta = \frac{\cos\theta_i(1+\beta^2) - 2\beta}{1 - 2\beta\cos\theta_i + \beta^2} \tag{4.2}$$

where $\beta = V/c$ and V is the velocity of the mirror, c is velocity of light and θ_i is the angle of incidence.

4.1.2 Statistics

Target correlation and fluctuation time

For an accelerating target the correlation function is approximated by [8]:

$$\begin{aligned} \rho &= \sin(\pi f_{\max} T_{ob})/(\pi f_{\max} T_{ob}) \\ f_{\max} &= 2a_r L_x/V_t\lambda \end{aligned} \tag{4.3}$$

89

where L_x is the target cross dimension to other radar line of sight, V_t is the target velocity, T_{ob} is the observation time and a_r is the target radial acceleration about its turning point.

For a nonaccelerating target, with a uniform distribution of contributors to the radar cross section [9]

$$\rho = \sin(\pi\delta/\theta)/(\pi\delta/\theta) \qquad (4.4)$$

and for a pair of reflectors contributing to the main RCS lobe, or for a view of a planar reflector in the far-off sidelobes:

$$\rho = \cos(\pi\delta/\theta) \qquad (4.5)$$

where δ is the angle by which the target axis rotates relative to the line of sight, during the observation time,

$$\delta = \frac{V_t T_{ob}}{R_t} \sin\alpha \qquad (4.6)$$

R_t is range to target

α is the angle between the radar line of sight and the target velocity vector

$\theta = \lambda/2D_{\text{eff}}$

$D_{\text{eff}} = \left[\left(\frac{L}{2}\sin\alpha\right)^2 + \left(\frac{W}{2}\cos\alpha\right)^2\right]^{0.5}$

L is the target length

W is the target wing span

When $\rho = 0.6$, $\qquad \delta = \theta/3$.

And for the nonaccelerating target the fluctuation time is [8]:

$$t_f = \lambda R_t/6D_{\text{eff}}V\sin\alpha \qquad (4.7)$$

Figure 4.1 shows the target fluctuation time for two target width-to-length ratios, 0.1 appropriate to missiles and 1 appropriate to airplanes. The data are for a target length to radar wavelength ratio of 100. These data scale inversely with other target length to wavelength ratios. A L/λ of 1000 will therefore decrease the fluctuation time values by 10. The data are given for a target velocity of 300 m/s and a range of 100 km. We see for the plotted conditions that the target has slow fluctuation (Swerling Case 1), i.e., it is correlated over normal radar integration time periods. The above aircraft example considered uniform target travel and not normal random motions in yaw, pitch, and roll. These random motions will reduce the correlation times. Howard, [1], pp. 18.34–18.46, discusses this additional target noise source. It may be possible for the reader to apply the preceding formulas to estimating the effects of random target motions.

Correlation between glint and fading

For all targets when the target signal is small due to fading the glint will be large. For ships [10] the correlation coefficient between glint and signal fading is -0.8.

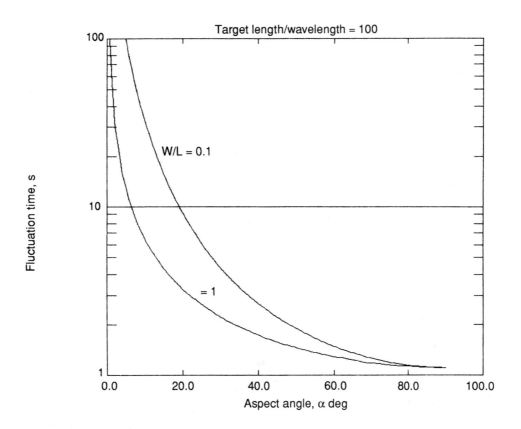

Figure 4.1: Illustration of aspect and target dependent fluctuation for a nonaccelerating target.

RCS estimation of complex target shapes

An often used approach consists of three steps:

1. The complex shape is decomposed into an ensemble of simple components. These components may be spheres, ogives, flat plates cylinders, wires, etc.

2. The RCS of each simple shape is calculated as a function of aspect angle from approximation formulas or graphical data.

3. The RCS for each simple shape at each aspect angle of interest is arithmetically added. This assumes random phase between the simple components.

An example target and the corresponding RCS estimate is shown in Figure 4.2. The lettered labels on the RCS curve show the predominant contribution to the RCS for each angular region.

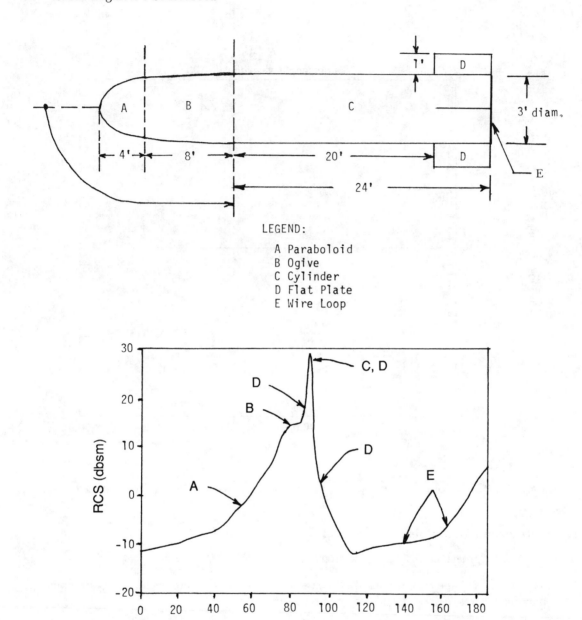

LEGEND:

A Paraboloid
B Ogive
C Cylinder
D Flat Plate
E Wire Loop

Figure 4.2: Example complex target shape and resulting estimated radar cross section, from [8].

4.2 RCS of Generic Shapes

4.2.1 Flat Surfaces

Flat plate

For an electrically large plate at normal incidence:

$$\sigma = \frac{4\pi A^2}{\lambda^2} \tag{4.8}$$

where A is the plate area.

For other angles, [11]:

$$\sigma(\theta, \phi) = \frac{4\pi A^2}{\lambda^2} \left[\frac{\sin(ka\sin\theta\cos\phi)}{ka\sin\theta\cos\phi} \frac{\sin(kb\sin\theta\sin\phi)}{kb\sin\theta\sin\phi} \right]^2 \cos^2\theta \tag{4.9}$$

where $k = 2\pi/\lambda$, θ is the angle relative to the normal to the plate and ϕ is the angle between the plane that contains the normal and the radar line-of-sight and the a dimension.

Circular disc

For an electrically large disc at normal incidence:

$$\sigma = \frac{4\pi A^2}{\lambda^2} \tag{4.10}$$

For other angles, [11]:

$$\sigma = \frac{4\pi A^2}{\lambda^2} 2 \left[\frac{J_1(2ka\sin\theta)}{2ka\sin\theta} \right]^2 \cos^2\theta \tag{4.11}$$

where J_1 is the first-order Bessel function.

For an electrically small disc where $L/\lambda << 1$ and $ka << 1$:

$$\sigma \approx L^2(ka)^4 \tag{4.12}$$

Corner reflector

For a corner reflector, consisting of two square plates, also known as a dihedral reflector, viewed in the plane normal to both plates, the RCS is [6]:

$$\sigma \approx \frac{16\pi a^4 \sin^4\theta}{\lambda^2}, 0 \leq \theta \leq 45^\circ$$
$$\text{use } (90 - \theta) \text{ for } 45^\circ \leq \theta \leq 90^\circ \tag{4.13}$$

where a is height and width of each plate of the dihedral, and θ is the angle relative to the bisector angle of the dihedral. σ has a maximum of $(8\pi a^4/\lambda^2)$ at $\theta = 45^\circ$.

For angles $\pm 75^\circ$ about the dihedral bisector the RCS is a combination of dihedral, flat plate and edge reflections.

$$\sigma = \sqrt{\sigma_{fp}^2 + \sigma_{di} + \sigma_e} \tag{4.14}$$

where σ_{fp} is the flat plate RCS, σ_{di} is the dihedral RCS, and σ_e is the edge RCS. See section 4.2.10 for wire and edge RCS values applicable to σ_e.

The predicted backscattered field from a dihedral reflector with $a/\lambda = 3.333$ is shown in Figure 4.3, [12]. The figure includes data for perfectly conducting and lossy reflectors for which the surface impedance, z is shown. The major lobes at 0 and 90° are flat plate contributions.

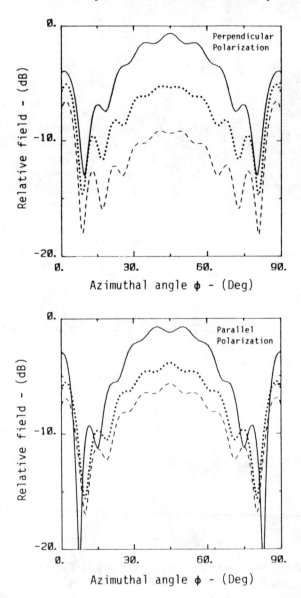

Figure 4.3: Backscatter field versus asimuth angle (45° is dihedral bisector angle) for perfectly conducting case (solid line), for $Z = 0.9 + j2.5$ (dotted line) and $Z = 1.98 + j2.43$ (dashed line), from [12].

Trihedral reflectors

We are considering three mutually perpendicular, centered and equal planes. Eight spaces with three sides each are so formed. Each intersecting side has length x. The RCS when viewed from the direction of maximum triple-bounce area, if the non-intersecting sides are square, will be [6]:

$$\sigma = 3\frac{4\pi x^4}{\lambda^2} \tag{4.15}$$

If the intersecting planes are circular:

$$\sigma = 1.202\frac{4\pi x^4}{\lambda^2} \tag{4.16}$$

If the non-intersecting sides form triangles rather than squares:

$$\sigma = \frac{1}{3}\frac{4\pi x^4}{\lambda^2} \tag{4.17}$$

4.2.2 Volumes

Small targets

In the Rayleigh region where a target is small relative to a wavelength, has a smooth surface, and does not deviate too much from a sphere, the RCS is given by [6]:

$$\sigma = \frac{4}{\pi}k^4 V^2 F^2 \tag{4.18}$$

where $k = 2\pi/\lambda$, V is the target volume, and $F \approx 1$ is a shape factor.

Sphere

Figure 4.4 shows the cross section of a sphere as a function of its size. The cross section values have been normalized by the projected area and the radius, a, has been normalized by $2\pi/\lambda$. In the region where the sphere is small relative to a wavelength and there are no oscillations of the RCS values (the Rayleigh region):

$$\sigma_R \approx 9\pi a^2 \left(\frac{2\pi a}{\lambda}\right)^4 \tag{4.19}$$

In the region where the sphere is large relative to a wavelength, the optics region:

$$\sigma_o \approx \pi a^2 \tag{4.20}$$

The region between the Rayleigh and optics regions, known as the resonant region, is where constructive and destructive interference occurs between specular reflected and creeping waves. The specular return is given by the optics region cross section. The creeping wave by [6]:

$$\sigma_c \approx \pi a^2 \left(\frac{2\pi a}{\lambda}\right)^{-5/2} \tag{4.21}$$

The maximum or minimum values of RCS in the oscillatory region are given by:

$$\sigma \approx \sqrt{\sigma_o^2 \pm \sigma_c^2} \tag{4.22}$$

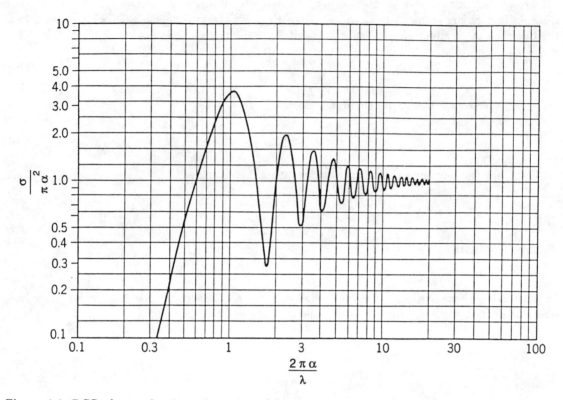

Figure 4.4: RCS of a conducting sphere, from [8].

4.2.3   Paraboloid

The head-on RCS is:

$$\sigma = \pi a^2 \tag{4.22}$$

where a is the radius of the apex curvature.

4.2.4 Cone

Infinite cone

The head-on RCS for an infinite cone is:

$$\sigma = \frac{\lambda^2}{16\pi} \tan^4 \alpha \qquad (4.24)$$

where α is the half-angle of the cone.

Blume and Kahl give the RCS for an elliptic cone as [13]:

$$\sigma = \frac{\lambda^2}{16\pi} \tan^2 \alpha_{\min} \tan^2 \alpha_{\max} \qquad (4.25)$$

In this instance $\alpha_{\min}$ is the half-cone angle of the elliptic minor axis and $\alpha_{\max}$ corresponds to the elliptic major axis.

The authors furthermore present relationships for the bistatic RCS:

$$\sigma(\beta) = \sigma(0) \left[\frac{2(1 + \cos 2\alpha_{\min/\max})^3}{(1 + \cos \beta)(\cos \beta + \cos 2\alpha_{\min/\max})^3} \right] \qquad (4.26)$$

where β is the bistatic angle, and $\sigma(\beta)$ is in the plane of the cone angle; either $\alpha_{\min}$ or $\alpha_{\max}$. β is restricted to $0 \leq \beta < \pi - 2\alpha_{\min/\max}$.

Finite cone

Choi et al. [14] present data for a finite cone for which $\alpha = 15^o$. Their data are shown in Figures 4.5, 4.6, and 4.7. In these figures a is the cone base radius, $k = 2\pi/\lambda$, and the RCS values are plotted as a function of the angle from the nose-on cone axis. The solid lines in each figure are the cited authors work. The remaining data are from other sources cited by Choi et al.

Finite cones blunted with a spherical nose were investigated by Blore, [2]. His nose-on RCS data for various apex radii (R_n) and base diameters for 8° half-cone angle are shown in Figure 4.8. Additional data for various aspect angles are given in Figure 4.9 and for various nose (R_n) and base (R_b) radii in Figure 4.10.

4.2.5 Cylinder

For a right circular cylinder with dimensions large relative to a wavelength for which the axis is perpendicular to the direction of propagation:

$$\sigma = \frac{2\pi a L^2}{\lambda} \qquad (4.27)$$

where a is the cylinder radius, and L is the length.

If the radar line-of-sight is not perpendicular to the axis, and if $(4\pi a \sin\theta)/\lambda >> 1$, [11]:

$$\sigma = \frac{a\lambda \sin\theta}{2\pi} \left\{ \frac{\sin[(2\pi L/\lambda) \cos\theta]}{\cos\theta} \right\}^2 \qquad (4.28)$$

where θ is the angle between the line-of-sight and the cylinder axis.

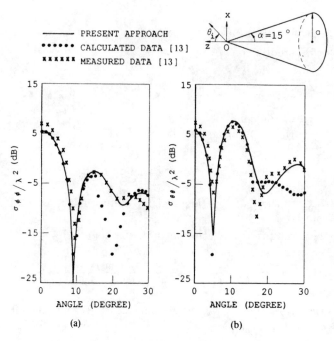

Figure 4.5: Backscattering RCS of a cone with $ka = 9.642$ and $\alpha = 15^o$ for (a) vertical polarization and (b) horizontal polarization, from [14].

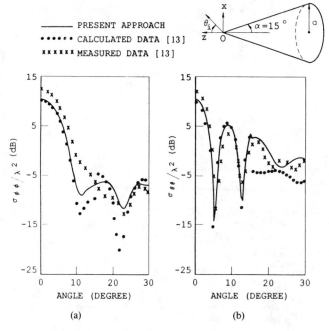

Figure 4.6: Backscattering RCS of a cone with $ka = 11.527$ and $\alpha = 15°$ for (a) vertical polarization and (b) horizontal polarization, from [14].

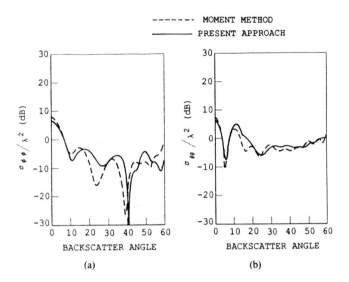

Figure 4.7: Backscattering RCS of a cone with $ka = 9.02$ and $\alpha = 15°$ for (a) vertical polarization and (b) horizontal polarization, from [14].

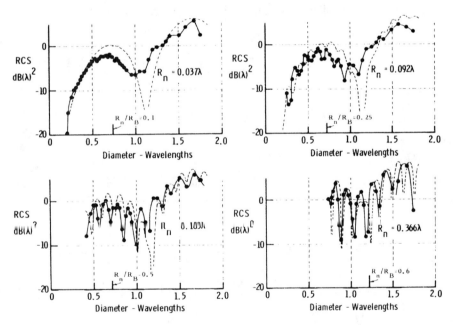

Figure 4.8: Nose-on backscatter from spherically blunted $8°$ finite cones, from [2].

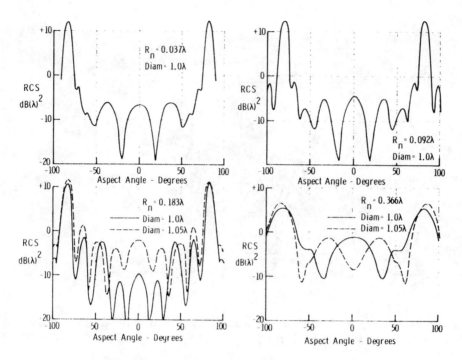

Figure 4.9: Backscatter from spherically blunted 8° finite cones as a function of aspect angle, from [2].

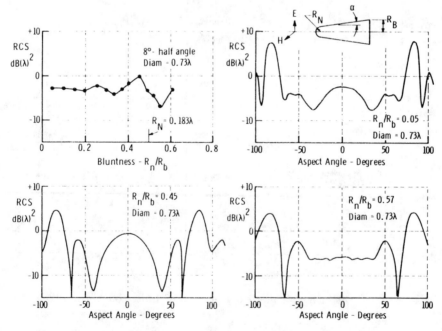

Figure 4.10: Backscatter from spherically blunted 8° finite cones for various nose and base radii, from [2].

4.2.6   Tubular Cylinder

The RCS of an open tubular cylinder, [15], is shown in Figure 4.11 for a cylinder with one end closed. The results from the work of Ling et al. are shown in Figures 4.12–4.16.

Tubular cylinder with blade termination

An open cylinder terminated at the opposite end with a rotating blade structure has a RCS characteristic as shown by Pathak and Burkholder [17] in Figure 4.17.

4.2.7   Tubular S-bend Cylinder

Ling et al. [16] predict the RCS for an S-bend tubular cylinder as shown in Figure 4.18. Their results are summarized in Figure 4.19.

4.2.8   Circular Ogive

The nose-on RCS of an electrically large circular ogive is the same as the infinite cone where in the ogive case the half-angle is determined by the tangent line at the ogive apex:

$$\sigma = \frac{\lambda^2}{16\pi} \tan^4 \delta \tag{4.29}$$

where δ is the half-angle formed by the ogive axis and the tangent line at the ogive apex.

4.2.9   Prolate Ellipsoid of Revolution

The RCS along the major axis of an electrically large ellipsoid of revolution is:

$$\sigma = \frac{\pi b^4}{a^2} \tag{4.30}$$

where a and b are the semimajor and semiminor axes.

4.2.10   One-Dimensional Objects

Wire

For normal incidence, $ka \ll 1$ and polarization parallel to the wire, [6]:

$$\sigma_\parallel \approx \frac{\pi L^2}{(\pi/2)^2 + [\ln(\lambda/1.78\pi a)]^2} \tag{4.31}$$

except for wire lengths near odd multiples of $\lambda/2$ where resonances occur. Some measurements at these wire length resonances indicate a cross section that is 50 to 100% larger than the approximating equation.

Similarly for polarization perpendicular to the wire but neglecting the comments regarding resonances:

$$\sigma_\perp \approx \frac{9}{4}\pi L^2 (ka)^4 \tag{4.32}$$

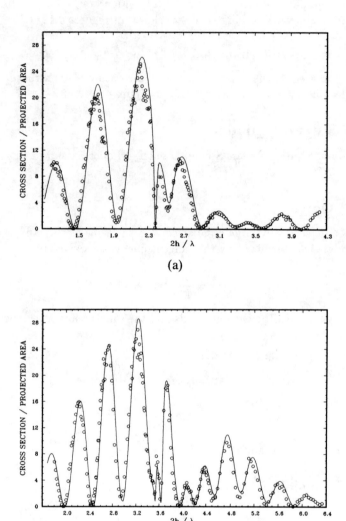

Figure 4.11: Backscatter along the axis of a tubular cylinder, (a) length/diameter= 4, (b) length/diameter= 5, theory (solid curve), experiment (circles), from [15].

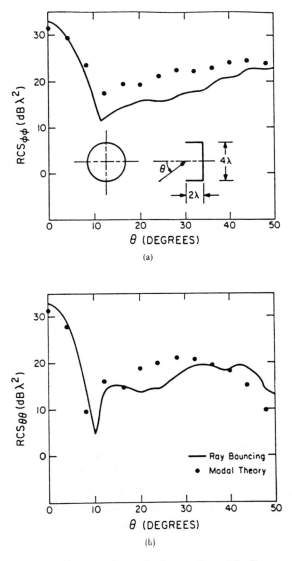

Figure 4.12: Backscatter from a right circular cylinder cavity with diameter 4λ and depth 2λ, (a) $\phi\phi$ polarization, (b) $\theta\theta$ polarization. From [16].

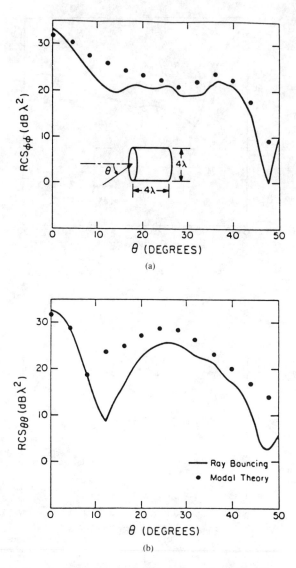

Figure 4.13: Backscatter from a right circular cylinder cavity with diameter 4λ and depth 4λ, (a) $\phi\phi$ polarization, (b) $\theta\theta$ polarization. From [16].

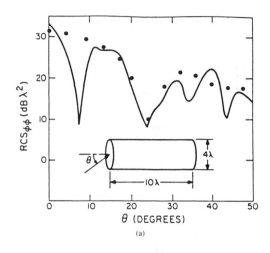

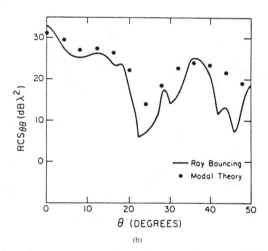

Figure 4.14: Backscatter from a right circular cylinder cavity with diameter 4λ and depth 10λ, (a) φφ polarization, (b) θθ polarization. From [16].

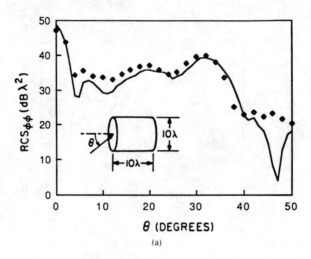

(a)

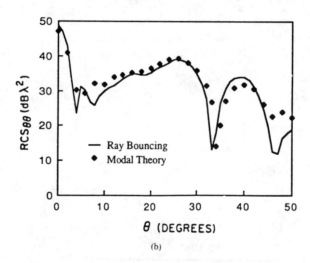

(b)

Figure 4.15: Backscatter from a right circular cylinder cavity with diameter 10λ and depth 10λ, (a) $\phi\phi$ polarization, (b) $\theta\theta$ polarization. From [16].

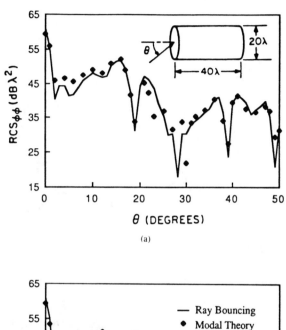

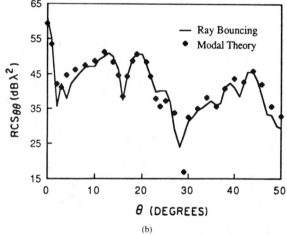

Figure 4.16: Backscatter from a right circular cylinder cavity with diameter 20λ and depth 40λ, (a) $\phi\phi$ polarization, (b) $\theta\theta$ polarization. From [16].

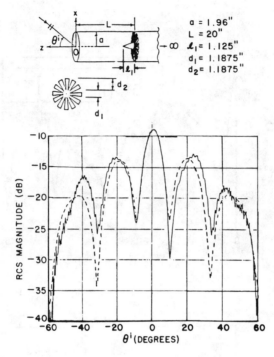

Figure 4.17: RCS of an open-ended right cylinder with a hub and blade termination at 10 GHz and vertical polarization. Solid line is measured and broken line is calculated, from [17].

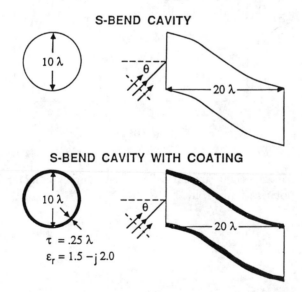

Figure 4.18: Geometry of the *S*-bend cavity, from [16].

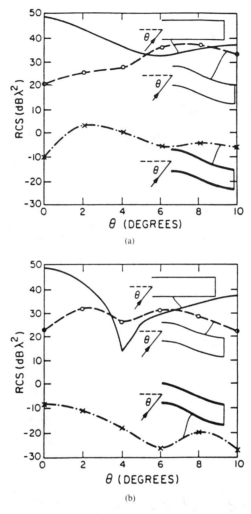

Figure 4.19: RCS of the uncoated and coated S-bends shown in Figure 4.18. (a) $\phi\phi$ polarization, (b) $\theta\theta$ polarization, from [16].

For a half-wave long wire or a quarter wave monopole viewed perpendicular to the line-of-sight and with parallel polarization:

$$\sigma = 0.86\lambda^2 \tag{4.33}$$

and for all aspects and polarizations the average value is:

$$\sigma = 0.17\lambda^2 \tag{4.34}$$

Edges

Knott [18] gives the RCS for a straight edge perpendicular to the line-of-sight as:

$$\sigma_e \approx L^2/\pi \tag{4.35}$$

and for a curved edge as:

$$\sigma_e \approx a\lambda/2 \tag{4.36}$$

where a is the radius of the edge contour.

4.3   RCS of Complex/Unique Shapes

4.3.1   Airplanes and Helicopters

Table 4.1 gives approximate nose-on RCS values for a variety of aircraft.

Another approximation given by Filippelli [19] for the nose-on aircraft RCS is:

$$\sigma \approx 0.01L^2 \tag{4.37}$$

where L is the aircraft length.

An aircraft has several principal echo regions that characterize the RCS signature. These regions are the nose ($\pm 60°$), broadside ($\pm 30°$), and tail ($\pm 60°$), [23]. Contributions to the RCS in the nose region are due to the radar compartment, canopy, engine inlets, and leading edges of the wings. Contributions to the RCS in the broadside region are the fuselage and vertical tail. Contributions to the RCS in the tail region are the engine exhaust nozzles and trailing edges of wings.

Typical fighter and bomber RCS dependence on microwave frequencies is shown in Figure 4.20.

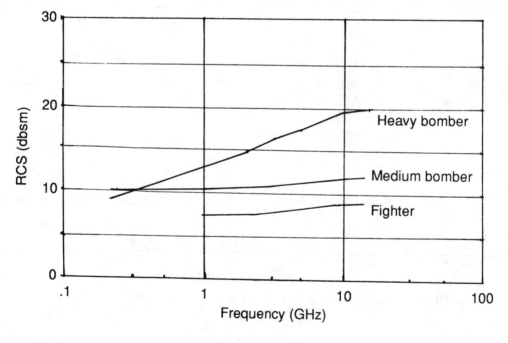

Figure 4.20: Example nose-on aircraft RCS variation with microwave frequencies, from [23].

Table 4.1: Nose-on RCS of aircraft. Values from a number of sources as indicated. Some values are thought to be estimates. (Table continued on next page.)

Designation	Name	RCS (m^2)	Ref
Aero 500	Aero Commander	6.5	[3]
B-1A	prototype	10	[20]
B-1B		1	[20]
B-26		100	[24]
B-29	Super Fortess	100	[3]
B-47	Stratojet	16	[3]
B-47		60	[24]
B-52	Strato Fortess	125	[3]
B-57B	Canberra	10	[3]
707	Boeing 707	32	[3]
707	at UHF	10	[22]
707	at L-band	25	[22]
707	at S-band	40	[22]
707	at C-band	50	[22]
707	at X-band	60	[22]
727	Boeing 727	10	[22]
727	(broadside)	125	[25]
720, 727	Boeing 720, 727	25	[3]
737	Boeing 737	3	[21]
737	Boeing 737	600	[23]
Britannia	Bristol Britannia	25	[3]
C-54		60	[24]
C-121	Constellation	100	[3]
C-130	Hercules	80	[3]
	Caravelle	16	[3]
	Comet	40	[3]
Cessna 180		1.5	[3]
Cessna 310B		4	[3]
Convair 240,340,440	Metropolitan	40	[3]

(Table continued from previous page.)

Designation	Name	RCS (m^2)	Ref
DC-3	Dakota	20	[3]
DC-8		32	[3]
DC-9		25	[3]
F-1/FJ	Sabre	5	[3]
F-4	Phantom	10	[3]
F-9F	Couger	12.5	[3]
F-27	Fokker Friendship	25	[3]
F-86	Sabre	5	[3]
F-104	Starfighter	5	[3]
IL-28	Beagle	8	[3]
	Javelin	8	[3]
	Lamps	25	[3]
L-1011	(broadside)	49	[25]
Mig-21	Fishbed	4	[3]
P-3A	Orion	80	[3]
P-3B	Orion	95	[3]
T-38		1	[24]
TU-16	Badger	25	[3]
TU-20	Bison	40	[3]
TU-95	Bear	125	[3]
	Viscount	16	[3]

Helicopters

A representative spectrum resulting from a short-term measurement of a helicopter is shown in Figure 4.21. Spectral lines would be evident if the helicopter wave observed over a prolonged period of time with a high-resolution doppler processor, changing the continuous spectrum into discrete lines at the blade rate.

The doppler shift associated with the tip velocity is:

$$f_{bl} = \frac{4\pi f_r L}{\lambda} \qquad (4.38)$$

where L is the blade length and f_r is the rotation rate in revolutions per second.

The number of spectral lines created across the width $2f_{bl}$ is, [23]:

$$N = \frac{2f_{bl}}{N_b f_r} \qquad (4.39)$$

where N_b is the number of blades.

The blade length can be determined from:

$$L = N N_b \frac{\lambda}{4\pi} \qquad (4.40)$$

048-PT OVERLAPPED HANNING FFTS

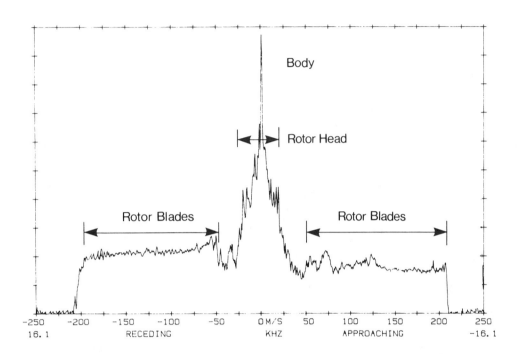

Figure 4.21: Spectral data from Hovering Helicopter. Note that the velocity scale is independent of wavelength, but that the doppler scale (shown reversed in sign) is for X-band. From [26].

4.3.2 Missiles

A measured polar plot of a generic missile is shown in Figure 4.22 to illustrate relative contributions of the nose, fuselage, wing and tail assemblies. The scale value of the outer circle is +30 dBm^2. The scale values decrease in 10 dB steps from the outer circle. We can see that the nose-on RCS is about 0.5 m^2 and the broadside RCS is about 100 m^2.

Youssef [28] presents additional data for another generic missile, Figure 4.23. Further data from Youssef, [29] shows that the nose-on RCS for his generic 1m long and 1/8 m diameter model varies with frequency, 4 to 17GHa, for vertical polarization as:

$$\sigma_{\text{nose}}^{V} \approx 0.7\lambda^{1.7} \tag{4.41}$$

and for horizontal polarization as:

$$\sigma_{\text{nose}}^{H} \approx 0.17\lambda^{1.3} \tag{4.42}$$

Another generic missile RCS characteristics is shown in Figure 4.24, from [23] that indicates the effect of radar frequency. The missle length is 4.2 m. The same missile shape yields bistatic RCS as shown in comparison with monostatic RCS in Figure 4.25.

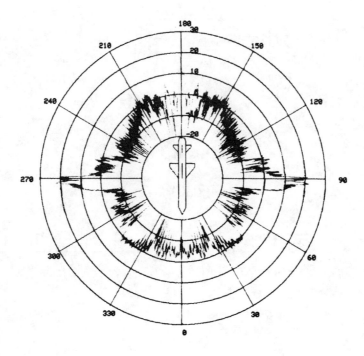

Figure 4.22: Polar plot of measured RCS of generic missile.

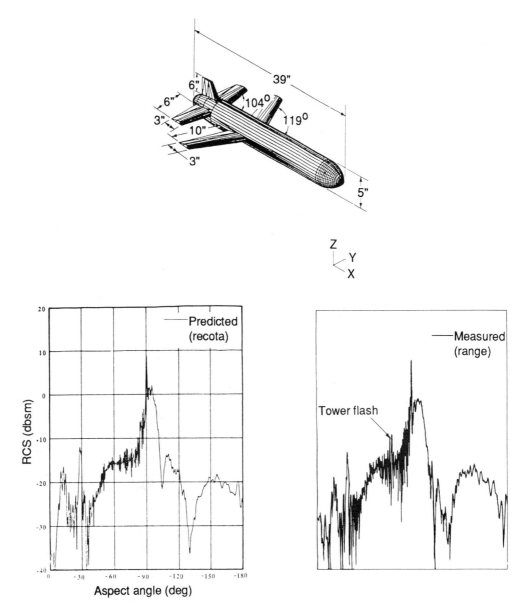

Figure 4.23: Predicted and measured RCS for a generic missile, from [28].

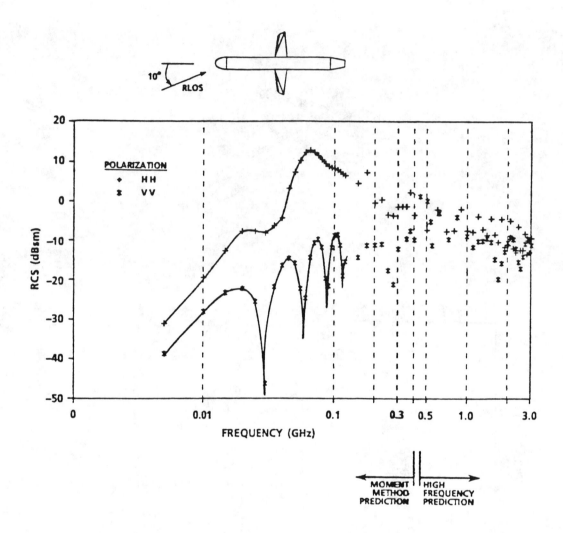

Figure 4.24: Simulated RCS versus frequency for a generic missile shape, from [8] after Cha et al, [30].

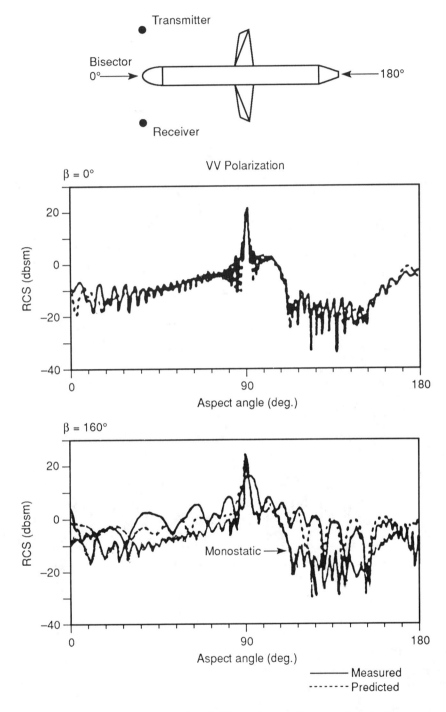

Figure 4.25: (a) Bistatic vertical polarization RCS prediction for generic missile shape as in Figure 4.24, with overlay of monostatic values, from [8] after Cha et al. [30].

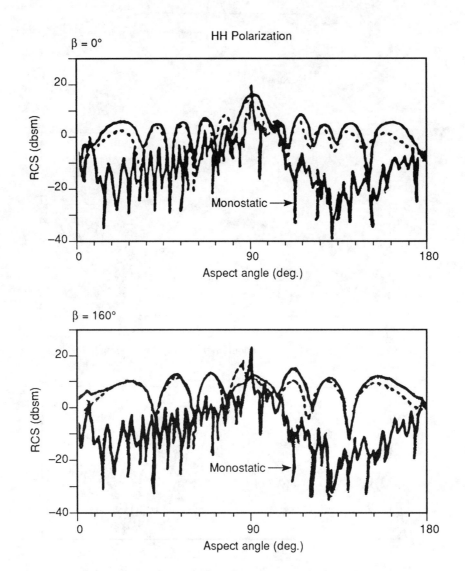

Figure 4.26: (b) Bistatic horizontal polarization RCS prediction for generic missile shape as in Figure 4.24, with overlay of monostatic values, from [8] after Cha et al. [30].

4.3.3 Maritime vessels

Skolnik [24] gives a generalized relationship for predicting the RCS of ships at elevation angles near grazing as:

$$\sigma \approx 1644 F_{GHz}^{1/2} D^{3/2} \tag{4.43}$$

where the cross section is in m^2, F_{GHz} is the radar frequency, GHz, and D is the ship displacement in kilotons. The approximation varies with aspect angle. Anderson [31] suggests adding 13 dB to approximate the maximum RCS and subtracting 8 dB to approximate the minimum RCS to expect for aspect angle dependency. Maximum RCS can be expected for broadside aspect angles and minimum RCS can be expected for bow or stern aspect angles. An approximation for displacement, [19] is:

$$D \approx 2.5 \times 10^{-6} L^3 \tag{4.44}$$

where L is the ship length, m.

For other than grazing angles, such as from airborne radar platforms:

$$\sigma \approx D \tag{4.45}$$

where σ is in units of m^2 and D in tons in this instance. Anderson gives a RCS weighting factor as a function of ships structure height, Figure 4.27, which may be applied to the above approximations to predict the vertical distribution of RCS. The integral of the distribution is 1 so there will be no change in the total RCS.

Generalized values of RCS for a variety of ships and references are given in Table 4.2.

4.3.4 Other Objects

Antennas

The RCS for antennas is based on the contribution of two components; the antenna mode, and the structure. The antenna mode component relates to coupling into the antenna feed whereas the

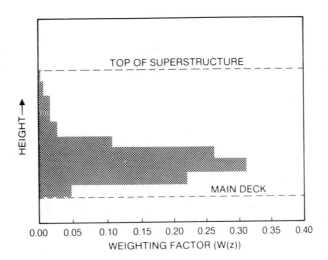

Figure 4.27: RCS weighting factor height distribution, from [31].

Table 4.2: Generalized maritime targets radar cross section values.

Target Type	Size(ft)	RCS(m²)	
Small open boat		0.02	[24]
Periscope		1	[32]
Small pleasure boat		2	[24]
Fishing vessel	20	6	[33]
Submarine snorkel		10	[32]
Cabin cruiser		10	[24]
Cabin cruiser	40	10	[23]
Wooden mine sweeper	144 × 25	80	[23]
Patrol boat	84	100	[23]
Small cargo ship		140	[23]
Trawler	525	150	[33]
Surfaced submarine	306 × 27	10-200	[23]
Freighter	1200	500	[33]
Freighter	2000	1000	[33]
Tanker		2230	[23]
Navy Picket Ship	441	200-3160	[23]
Loaded tanker		630-2130	[23]
Loaded freighter		7400	[23]
Navy cruiser	664	14000	[23]
Cargo ship	278 × 40	$100\text{-}10^6$	[23]
Coast Guard 3rd class		65	[23]
Coast Guard 8 × 26		370	[23]
Coast Guard 1st class		670	[23]

structural component is not related to coupling into the feed system. The antenna mode component is zero if the antenna is perfectly matched. Otherwise the antenna RCS is [35]:

$$\sigma = \frac{\lambda^2 G^2}{4\pi}\Gamma \tag{4.46}$$

where G is the antenna gain, and Γ is the power reflection coefficient at the antenna feed.

Hansen [34] makes the point that the phase relationships between the structural and antenna modes are important:

$$\sigma = \mid \sqrt{\sigma_s} - (1 - \Gamma)\sqrt{\sigma_a}\exp\phi_{\text{rel}} \mid \tag{4.47}$$

where σ_s is the structural mode RCS, σ_a is the antenna mode RCS, and ϕ_{rel} is the relative phase between the the terms of the relationship. With careful design attention to the real and imaginary parts of Γ, one may be able to obtain zero or near-zero scatter from an antenna. For many antennas, the structural scatter is about four times the antenna mode scatter.

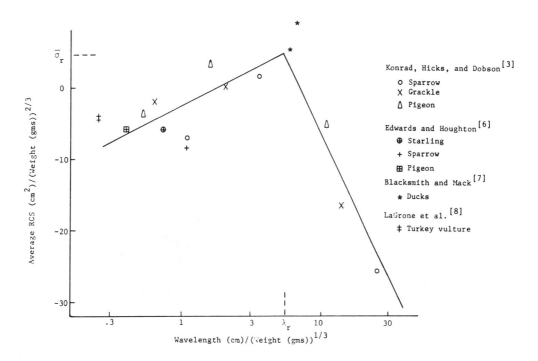

Figure 4.28: Normalized bird RCS, from [36].

Birds and insects

A general equation for the RCS of birds and insects is [8]:

$$\sigma_b \approx -46 + 5.8 \log W_b, \ \text{dB m}^2 \tag{4.48}$$

where W_b is the weight in grams of a bird or insect. Data indicate that there is a 5.5 dB standard deviation about the values predicted with the approximation. Pollon, [36] summarizes the frequency dependence of bird RCS with Figure 4.28. The wavelength is normalized by bird weight, $gm^{1/3}$, and the RCS in cm^2 is normalized by bird weight, $gm^{2/3}$.

Humans

The measured RCS, [37] of a 200 pound 6 ft man is shown in Figure 4.29. The top pair of curves are an average of RCS for front and rear aspect angles. The bottom pair of curves are for a side aspect angle. The top curve of each pair is for vertical polarization, the other for horizontal polarization. The RCS was found by measurement of other men to vary approximately in proportion to the weight of the man.

Miscellaneous

The general RCS of other objects is:

Pickup truck	200 [24]
Automobile	100
Bicycle	2

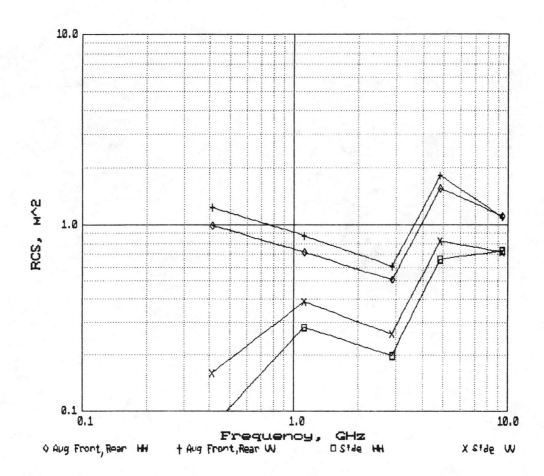

Figure 4.29: RCS of 200-pound 6-ft man, upper pair of curves are average of front and rear aspect, lower pair of curves are side aspect; +, x are vertical polarization; ◊, □ are horizontal polarization; after [37].

4.4   Low RCS Technology

4.4.1   Surface Wave Absorbing Materials

A thin layer of ferrite and synthetic rubber paint now (at time of reference) provide 20 to 30 dB of attenuation with 5 to 10 GHz bandwidth, [38].

Kumar [39] cautions that the proportion of carbon used in acetylene black rubber is important in achieving uniform reduction with material thickness to wavelength ratio. He found that increasing the proportion of carbon from 30% to 40% flattened out the peaks and valleys of reflectivity versus frequency for material thickness-to-wavelength ratios greater than about 0.3.

Broadband absorbers can be made with multiple layers of resistive sheets and low dielectric constant spacers (Jaumann absorber) Measured reflectivity of a Jaumann absorber is shown in Figure 4.30.

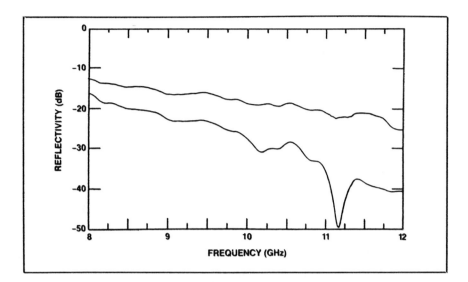

Figure 4.30: Reflectivity of three-layer Jaumann absorber. The two curves are from two measurements taken 10 cm apart, near the center of a 61 cm square panel, from [40].

4.4.2 Structural Absorbing Materials

According to Nengjing and Li [38] the considerations for achieving structural absorption are:

1. Use of composites of absorbing materials and nonmetallic structure.

   Achieves 10 dB attenuation at cm wavelengths

2. Use combination of filaments of wave absorbing materials and metallic or non-metallic structure combinations.

3. Replace original structure with the above item.

   Achieves 10 dB attenuation.

4. Use impedance loading of craft cavities.

   Design apertures of cavities so that diffraction fields of craft are reduced by cancellations produced by the loading field. This is effective when the radar wavelength is the order of the craft dimensions.

Bibliography

[1] Howard, D.D., Ch. 18 "Tracking Radar," Skolnik, M. (ed.) *Radar Handbook*, McGraw-Hill, New York, 1990.

[2] Blore, W.E., "The radar cross section of spherically blunted 8° right-circular cones," IEEE Trans. AP, March 1973.

[3] Rohan, P., *Surveillance Radar Performance Prediction*, Peter Peregrinus Ltd., London, 1983.

[4] Kelley, J.B., from personal notes, circa 1984.

[5] IEEE Standard Radar Definitions, std. 686–1977.

[6] Bodhar, D.G., Cairman IEEE Antenna Standards Committee, *personal communications*, February 9, 1990.

[7] Harfoush, F., A. Taflove, G. Kriegsmann, "A numerical technique of analyzing electromagnetic wave scattering from moving surfaces in one and two dimensions," IEEE Trans. AP-37, no. 1, Jan. 1989, pp.55-63.

[8] Morchin, W.C., *Airborne Early Warning Radar*, Artech House, Norwood, MA, 1990.

[9] Hanle, E., "Cross section variations with different target classes and their influences to detection and tracking," Int. Radar Conf., Paris, Apr. 24-28, 1989, pp. 135-140.

[10] Wallin, C. "Experimental investigation of correlatioinn between fading and glint for ship targets," Radar Conf. Paris 1989, Apr. 24-28.

[11] Blake, L.V., *Radar Range-Performance Analysis*, Artech House, Norwood, MA, 1986, p. 105.

[12] Corona, P., G. Ferrara, C. Gennarelli, "Backscattering by loaded and unloaded dihedral corners," IEEE Trans. AP-35, no. 10, Oct. 1987, pp. 1148-1153.

[13] Blume, S., G. Kahl, "The physical optics radar cross section of an elliptic cone," IEEE Trans. AP-35, no. 4, April 1987, pp. 457-460.

[14] Choi, J., N. Wang, L. Peters, P. Levy, "Near axial backscatytering from finite cones," IEEE Trans. AP-38, no. 8, Aug. 1990, pp. 1264-1272.

[15] Lee, H-M., "Electromatnetic scattering of tubular cylindrical structure-double series formulation and some results," IEEE Trans. AP-35, no. 4, Apr. 1987, pp. 384-390.

[16] Ling, H., R-C. Chou, S-W. Lee, "Shooting and bouncing rays: calculating the RCS of an arbitrarily shaped cavity," IEEE Trans. AP-37, no. 2, Feb. 1989, pp. 194-205.

[17] Pathak, P.H., R.J. Burkholder, "Modal, ray, and beam techniques for analyzing the EM scattering by open-ended waveguide cavities," IEEE Trans. AP-37, no. 5, May 1989, pp. 635-647.

[18] Knott, E.F., "Radar cross section," Ch. 11 of *Radar Handbook*, M. Skolnik, ed., 2nd edition, McGraw-Hill, 1990.

[19] Filippelli, L. et al. "A generic space based radar model with resource scheduler," NRL report 5059, AD 126639, April 1983.

[20] Adam, J.A., "How to design an invisible aircraft," IEEE Spectrum, April 1988.

[21] Knott, E.F., J.F. Shaeffer, M.T. Tuley, *Radar Cross Section*, Artech House, Norwood MA, 1985, p. 181.

[22] Williams, F.C., From personal notes, 1985.

[23] Chiavetta, R.J., "Target scattering fundamentals" Ch.2 of *Airborne Earling Warning Radar*, W. Morchin, ed., Artech House, Norwood, MA 1990, p. 77.

[24] Skolnik, M., *Introduction to Radar Systems*, McGraw-Hill 1980.

[25] Steinberg, B.D., D.L. Carlson, W. Lee, "Experimental localized radar cross sections of aircraft," Proc. IEEE, vol. 77, no. 5, 1989, pp. 663-669.

[26] MacKenzie, J.D., K. Dyer, K.D. Ward, "The measurement of radar cross section," Military Microwaves, MM86, June 24-26, 1986, Brighton, England.

[27] Thorn-EMI "Radar target modeling facilities," brochure, 1987.

[28] Youssef, N.N., "Radar cross section of complex targets," Proc.IEEE, vol. 77, no. 5, May 1989, pp. 722-734.

[29] Youssef, N.N., "RECOTA verification using scale model missile measurements," Boeing Aerospace memorandum 2-3745-ATD-145, Nov. 23, 1987.

[30] Cha, C., J. Michels, and E. Starczewski, "An RCS analysis of generic airborne vehicles," IEEE National Radar Conf. Proc. 1988, pp. 214-219.

[31] Anderson, K.D., "Radar measurements at 16.5 GHz in the Oceanic evaporation duct," IEEE Trans. AP-37, no. 1, Jan. 1989, pp. 100-106.

[32] "Equipment specification for AN/APS-137," Texas Instruments, Mar. 1982.

[33] "AN/APS-128," *Jane's Avionics*, 1982-83.

[34] Hansen, R.C., "Relationships between antennas as scatterers and as radiators," Proc. IEEE, vol. 77, no. 5, May 1989, pp. 659-662.

[35] Flowers, D., C. Ryan, "ECM transmitters," Microwave Journal, Feb. 1987, pp. 141-152.

[36] Pollon, G.E., "Distributions of radar angels," IEEE Trans. AES-8, no. 6, Nov. 1972, pp. 721-727.

[37] Schultz, F.V., R.C. Burgener, S. King, "Measurement of the radar cross section of a man," Proc. IRE, vol. 46, Feb. 1958, pp. 476-482.

[38] Nengjing, Li, "Radar ECCMs new area: anti-stealth and antiarm," DTIC AD-A188 986, FTD-ID(RS)T-1297-87, Feb. 16, 1988, approved for public release - distribution unlimited.

[39] Kumar, A., "Acetylene black rubber reduces target RCS," Microwaves & RF, Mar. 1987, pp. 85-86.

[40] Baker, D.E., C.A. van der Neut, "Reflection measurements of microwave absorbers," Microwave Journal, Dec. 1988, pp. 95-104.

Notes

Notes

Chapter 5

Detection

5.1 General Formulas

5.1.1 Evaluation of Error Function Integral

Helstrom [1] presents an approximating function for the Q function. From his results, for $x \geq 3$ one can use the zero-order approximation for which the error will be $< 0.4\%$:

$$
\begin{aligned}
Q(x) &\approx Q^{(0)}(x) = [2\pi(1 + \lambda_0^2)]^{1/2} \exp(1 - 0.5\lambda_0^2) \\
\lambda_0 &= -0.5[x + (x^2 + 4)^{1/2}]
\end{aligned}
\tag{5.1}
$$

and for $x \geq 1$ one can use the third-order approximation for which the error will be $< 0.4\%$:

$$
\begin{aligned}
Q(x) &\approx Q^{(3)} = Q^{(0)}(x)[1 - (3t^2/4)/(1 + 40t/9)]^{-1} \\
t &= 1/(1 + \lambda_0^2)
\end{aligned}
\tag{5.2}
$$
$$
\text{when } x >> 1 \lambda_0 \approx -x
$$

To find the inverse, $x = Q^{-1}(P)$, where $P = Q(x)$ from Urkowitz [3] for $Q(x) \leq 0.5$:

$$
\begin{aligned}
x &= g - \frac{2.515517 + 0.802853g + 0.010328g^2}{1 + 1.432788g + 0.189269g^2 + 0.001308g^3} \\
\text{where} \\
g &= \sqrt{-2\ln Q(x)} \text{ for } Q(x) \leq 0.5
\end{aligned}
\tag{5.3}
$$

To find the inverse, $x = Q^{-1}(P)$, for $Q(x) > 0.5$:

$$
\begin{aligned}
x &= -h + \frac{2.515517 + 0.802853h + 0.010328h^2}{1 + 1.432788h + 0.189269h^2 + 0.001308h^3} \\
\text{where} \\
h &= \sqrt{-2\ln(1 - Q(x))} \text{ for } Q(x) > 0.5
\end{aligned}
\tag{5.4}
$$

See Barton, [2] tables 2.1 and 2.2, for equations for coherent and noncoherent detection using the Q function.

5.1.2   Thresholding

Determination of threshold for noncoherent integration following square law detection

For $P_{fa} < 0.5$ the threshold voltage, using equation 5.3, is:

$$V_T = \sqrt{2\sigma^2/Y_b}$$
$$Y_b \approx N + \sqrt{N}\frac{-\ln P_{fa} - 1 - x}{(1.1N - 0.1)^{0.51}} + x\sqrt{N} \tag{5.5}$$

where x is found from $Q^{-1}(P)$ with:

$$g = \sqrt{-2\ln P_{fa}} \tag{5.6}$$

where σ is the rms noise and N is the number of integrated pulses.

An approximation for determining the detector threshold using CFAR in a heterogenous Rayleigh clutter, [4] is:

$$T_n = k^2\sigma_T^2$$
$$k^2 = (P_{fa}^{-1/n} - 1)n \tag{5.7}$$

where n is the number of samples which are estimated to be in the relevant clutter field, $n < K$, where K is the CFAR window size, and σ_T is the rms noise at the threshold.

Number of independent pulses

Kanter [5] gives the number of independent pulses of an exponentially correlated signal from a Rayleigh target in Gaussian noise as:

$$N_I = \frac{N}{1 + [2\rho^2/(1-\rho^2)][1 - N^{-1}(1-\rho^{2N})/(1-\rho^2)]} \tag{5.8}$$

where ρ is the correlation coefficient between successive pulses. Figure 5.1 presents representative results. For the Swerling 2 target the pulses are independent from pulse to pulse for which case the correlation coefficient is zero. For a Swerling 1 target the pulses are dependent from pulse to pulse for which case the correlation coefficient is 1.

The correlation time is :

$$t_c = T_{ob}/\ln(1/\rho) \tag{5.9}$$

where T_{ob} is the observation time of the N pulses.

5.1.3   Required Signal-to-Noise Ratio

Nonfluctuating target

From Tufts and Cann [8] after Albersheim for envelope detection in narrowband noise:

$$S/N = -5\log M + (6.2 + \frac{4.54}{\sqrt{M + 0.44}})\log(A + 0.12AB + 1.7B)\,,\text{dB} \tag{5.10}$$

where
$A = \ln(0.62/P_{fa})$
$B = \ln[P_d/(1 - P_d)]$
and where M is the number of independent samples, and P_d is the probability of detection.

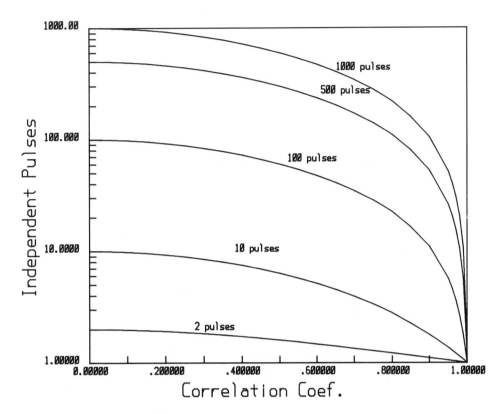

Figure 5.1: Number of independent pulses as a function of the correlation coefficient of N pulses, from [6].

Fluctuating target

Swerling 1 target. For noncoherent integration of N_e pulses for a Swerling 1 target from Jacobson [7]:

$$(S/N)_1 \approx \frac{1}{2\sqrt{N_e}} \left[\frac{\ln P_{fa}}{\ln P_d} - 1 \right] \qquad (5.11)$$

where $(S/N)_1$ is the average signal-to-noise required for one pulse and $10 \leq N_e \leq 100$, $10^{-n} \leq P_{fa} \leq 10^{-6}$, $.02 \leq P_d \leq .98$. For coherent integration or a single pulse the quantity in brackets is the exact value of $(S/N)_1$ of a Swerling case 1 target signal.

Swerling 2 target. After Urkowitz, [3], for $P_d > 0.5$:

$$S/N = -1 + \frac{Y_b}{N + \sqrt{N} \left[\frac{-\ln P_d - 1 - x}{(1.1N - .1)^{0.51}} + x \right]} \qquad (5.12)$$

where Y_b is found from equation 5.9 using P_d in the place of P_{fa} and,

$$h = \sqrt{-2\ln(1 - P_d)} \text{ in equation 5.4 to solve for } x$$

For $P_d \leq 0.5$ use $g = \sqrt{-2 \ln P_d}$ in equation 5.3 to solve for x in the above equation for S/N.

Swerling 3 target.   For noncoherent integration of N_e pulses for a Swerling 3 target:

$$(S/N)_1 \approx \frac{1}{\sqrt{N_e}} \left[\frac{\ln P_{fa}}{\ln P_d} - 1 \right]^{0.65} \tag{5.13}$$

for conditions of $10 \leq N_e \leq 1000$, $10^{-3} \leq P_{fa} \leq 10^{-6}$ and $0.1 \leq P_d \leq 0.9$. where $(S/N)_1$ is the signal-to-noise required for one pulse.

5.1.4 m Detections Out of n Successive Trials

For a single event probability, P, either P_d or P_{fa}, the solution for the output probability (P_o), after [10], is shown in table 5.1. It is assumed P is equal for each event.

Table 5.1: Probability of radar output probability for m successful detections out of n trials.

Detection Scheme	Output Probability, $P_o =$
2 out of 2	P^2
2 out of 3	$3P^2 - 2P^3$
3 out of 3	P^3
2 out of 4	$6P^2 + 4P^3 - 9P^4$
3 out of 4	$4P^3 - 3P^4$
4 out of 4	P^4

5.2   Cumulative Probability of Detection

5.2.1   Probability of Detecting on or Before a Range

For a Swerling 1 target the probability of detecting a target on or before a range r is, [11]:

$$P_{\text{cum}} \approx 1 - \exp\{-R_1/\Delta\} \exp[1.38 - 3.05(r/R_1) - (r/R_1)^{3.72}]\} \tag{5.14}$$

where $\Delta = V_t T_r$, V_t is the target radial velocity, T_r is the revisit time, R_1 is a scale range for which $P_d = 0.37$, [12], p.483.

5.2.2   Normalized P_{cum} Curves for Search

$$P_{\text{cum}} \approx 1 - \exp\left\{ -\left[\frac{(\Delta R/R)}{B} \right]^{1/A} \right\} \tag{5.15}$$

where the constants A and B are given in the following table for a maximum ratio $\Delta R/R$ for which the radar resources will be minimized.

Table 5.2: Table of constants for equation 5.15.

Swerling target	A	B
1	−0.84	0.37
3	−0.8	4.43

5.3 Reference Curves for Detection Probability

5.3.1 Nonfluctuating and Rayleigh Targets

Tables 5.3 and 5.4 refer to a compilation of figures that pertain to nonfluctuating and Rayleigh targets that have various degrees of fluctuation and correlation. The Swerling 1 and 3 targets referred to in the figures are completely correlated from pulse to pulse. The Swerling 2 target referred to in the figures is completly uncorrelated from pulse to pulse. The effect of partial correlation is shown in Table 5.4.

Table 5.3: Detection of Rayleigh and nonfluctuating targets in Gaussian noise.

Figure	Plot	No. of pulses N	P_d	Other parameters
Figure 5.2	P_d vs P_{fa}	1		nonfluctuating target, $E/N = -10$ to 15 dB
Figure 5.3	P_d vs P_{fa}	1		Swerling 1 target, $E/N = -5$ to 35
Figure 5.4	E/N vs N	1 to 200	0.75	nonfluctuating target, $P_{fa} = 10^{-2}$ to -10
Figure 5.5	E/N vs N	1 to 200	0.75	Swerling 1 target
Figure 5.6	E/N vs N	1 to 200	0.75	Swerling 2 target
Figure 5.7	E/N vs N	1 to 200	0.5	nonfluctuating target, $P_{fa} = 10^{-2}$ to -10
Figure 5.8	E/N vs N	1 to 200	0.6	nonfluctuating target, $P_{fa} = 10^{-2}$ to -10
Figure 5.9	E/N vs N	1 to 200	0.7	nonfluctuating target, $P_{fa} = 10^{-2}$ to -10
Figure 5.10	E/N vs N	1 to 200	0.75	nonfluctuating target, $P_{fa} = 10^{-2}$ to -10
Figure 5.11	E/N vs N	1 to 200	0.8	nonfluctuating target, $P_{fa} = 10^{-2}$ to -10
Figure 5.12	E/N vs N	1 to 200	0.9	nonfluctuating target, $P_{fa} = 10^{-2}$ to -10
Figure 5.13	E/N vs N	1 to 200	0.95	nonfluctuating target, $P_{fa} = 10^{-2}$ to -10
Figure 5.14	E/N vs N	1 to 200	0.5	Swerling 1 target, $P_{fa} = 10^{-2}$ to -10
Figure 5.15	E/N vs N	1 to 200	0.6	Swerling 1 target, $P_{fa} = 10^{-2}$ to -10
Figure 5.16	E/N vs N	1 to 200	0.7	Swerling 1 target, $P_{fa} = 10^{-2}$ to -10
Figure 5.17	E/N vs N	1 to 200	0.75	Swerling 1 target, $P_{fa} = 10^{-2}$ to -10
Figure 5.18	E/N vs N	1 to 200	0.8	Swerling 1 target, $P_{fa} = 10^{-2}$ to -10
Figure 5.19	E/N vs N	1 to 200	0.9	Swerling 1 target, $P_{fa} = 10^{-2}$ to -10
Figure 5.20	E/N vs N	1 to 200	0.95	Swerling 1 target, $P_{fa} = 10^{-2}$ to -10
Figure 5.21	E/N vs N	1 to 200	0.5	Swerling 2 target, $P_{fa} = 10^{-2}$ to -10
Figure 5.22	E/N vs N	1 to 200	0.6	Swerling 2 target, $P_{fa} = 10^{-2}$ to -10
Figure 5.23	E/N vs N	1 to 200	0.7	Swerling 2 target, $P_{fa} = 10^{-2}$ to -10
Figure 5.24	E/N vs N	1 to 200	0.75	Swerling 2 target, $P_{fa} = 10^{-2}$ to -10
Figure 5.25	E/N vs N	1 to 200	0.8	Swerling 2 target, $P_{fa} = 10^{-2}$ to -10
Figure 5.26	E/N vs N	1 to 200	0.9	Swerling 2 target, $P_{fa} = 10^{-2}$ to -10
Figure 5.27	E/N vs N	1 to 200	0.95	Swerling 2 target, $P_{fa} = 10^{-2}$ to -10

Table 5.4: Detection probability of targets having different
correlation coefficients in noise and clutter.

Figure	Plot	Tgt. correl. ρ_s	No. of pulses N	P_{fa}	Other parameters
Figure 5.28	P_d vs S/N	0	10	10^{-6}	Rayleigh clutter, Swerling 2 target
Figure 5.64	P_d vs S/N	0	20	10^{-6}	Rayleigh clutter, Swerling 2 target
Figure 5.30	P_d vs S/N	.1-.99	2	10^{-4}	
					Gaussian shaped autocorrelation function of the target signal for Figures 5.30 through 5.36
Figure 5.31	P_d vs S/N	.1-.99	2	10^{-5}	
Figure 5.32	P_d vs S/N	.1-.99	5	10^{-4}	
Figure 5.33	P_d vs S/N	.1-.99	5	10^{-5}	
Figure 5.34	P_d vs S/N	.1-.99	10	10^{-4}	
Figure 5.35	P_d vs S/N	.1-.99	10	10^{-5}	
Figure 5.36	P_d vs S/N	.1-.99	2	10^{-6}	
Figure 5.37	P_d vs S/N	0.8	5	10^{-4}	Comparison of exponential ACF with Gaussian
Figure 5.38	P_d vs S/N	0-.99	2	10^{-4}	$\rho_c = .9$, $CNR = 20$ dB
Figure 5.39	P_d vs S/N	0-.99	2	10^{-4}	$\rho_c = .99$, $CNR = 20$dB
Figure 5.40	P_d vs S/N	0-.99	2	10^{-5}	$\rho_c = .9$, $CNR = 20$ dB
Figure 5.41	P_d vs S/N	0-0.8	5	10^{-4}	$\rho_c = .5$, $CNR = 20$ dB
Figure 5.42	P_d vs S/N	0-0.8	5	10^{-5}	$\rho_c = .5$, $CNR = 20$ dB
Figure 5.43	P_d vs S/N	0-0.99	2	10^{-6}	$\rho_c = .5$, $CNR = 20$ dB
Figure 5.44	P_d vs S/N	0.1 - 0.99	2	10^{-4}	$\rho_c = .9$, $CNR = 20$ dB
Figure 5.45	P_d vs S/N	0.1,.5,.8	5	10^{-4}	$\rho_c = .9$, $CNR = 20$ dB
Figure 5.46	Det.Loss	0.5,.8	2	10^{-4}	$\rho_c = .5$, $CNR = 20$ dB, Detection comparison with and without estimation
Figure 5.47	Det.Loss	0.5,.8	2	10^{-4}	$\rho_c = .9$, $CNR = 20$ dB, 5-,9-,12-cell CFAR

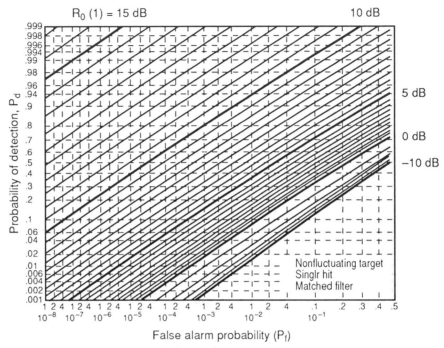

Figure 5.2: Optimum coherent radar detection performance, nonfluctuating target, from [13].

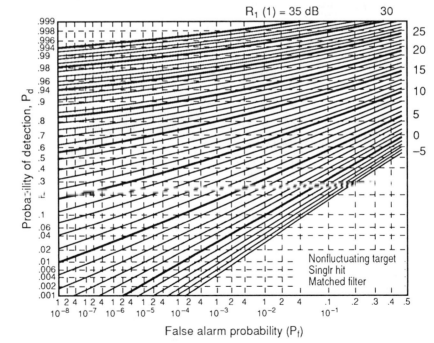

Figure 5.3: Optimum coherent radar detection performance, Swerling 1 target, from [13].

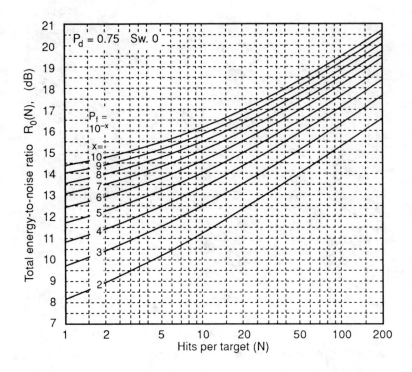

Figure 5.4: Optimum noncoherent radar detection performance, nonfluctuating target, from [13].

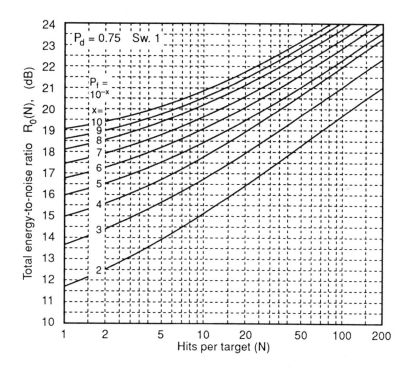

Figure 5.5: Optimum noncoherent radar detection performance, Swerling 1 target, from [13].

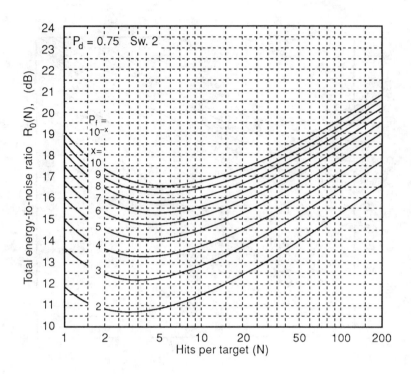

Figure 5.6: Optimum noncoherent radar detection performance, Swerling 2 target, from [13].

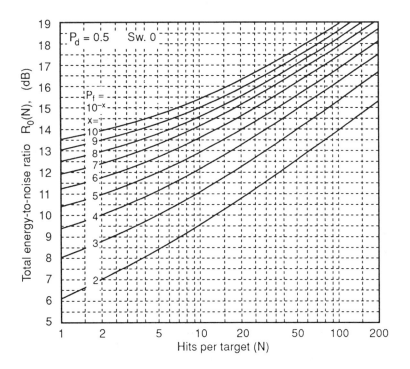

Figure 5.7: Square-law noncoherent detection performance, nonfluctuating target, $P_d = 0.5$, from [13].

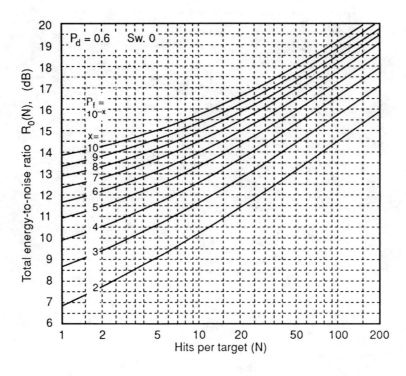

Figure 5.8: Square-law noncoherent detection performance, nonfluctuating target, $P_d = 0.6$, from [13].

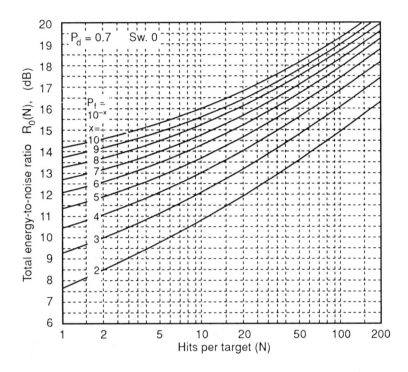

Figure 5.9: Square-law noncoherent detection performance, nonfluctuating target, $P_d = 0.7$, from [13].

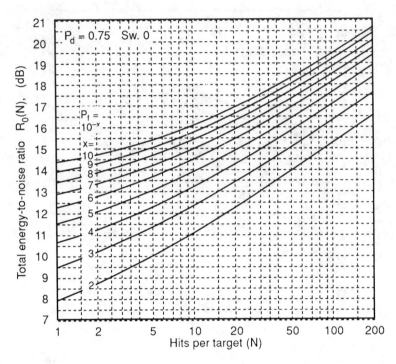

Figure 5.10: Square-law noncoherent detection performance, nonfluctuating target, $P_d = 0.75$, from [13].

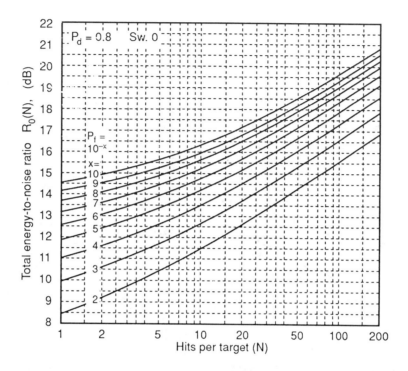

Figure 5.11: Square-law noncoherent detection performance, nonfluctuating target, $P_d = 0.8$, from [13].

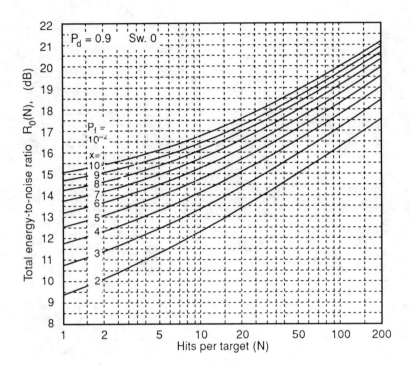

Figure 5.12: Square-law noncoherent detection performance, nonfluctuating target, $P_d = 0.9$, from [13].

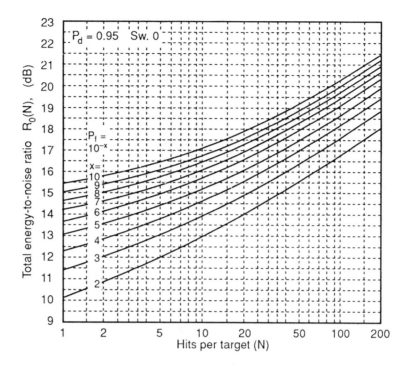

Figure 5.13: Square-law noncoherent detection performance, nonfluctuating target, $P_d = 0.95$, from [13].

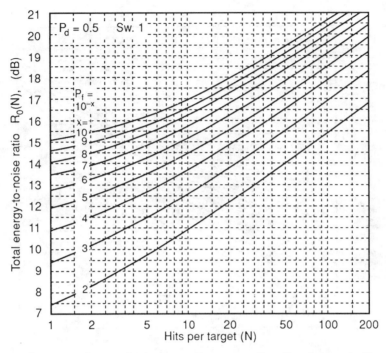

Figure 5.14: Square-law noncoherent detection performance, Swerling 1 target, $P_d = 0.5$, from [13].

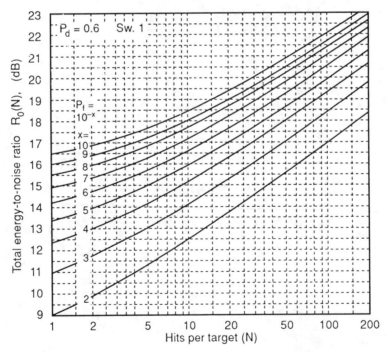

Figure 5.15: Square-law noncoherent detection performance, Swerling 1 target, $P_d = 0.6$, from [13].

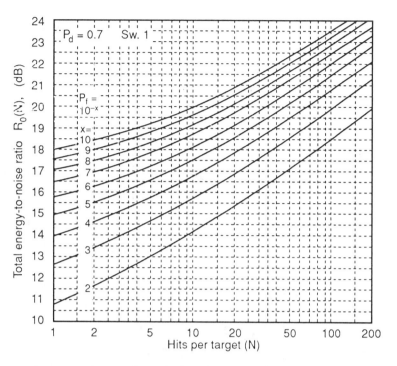

Figure 5.16: Square-law noncoherent detection performance, Swerling 1 target, $P_d = 0.7$, from [13].

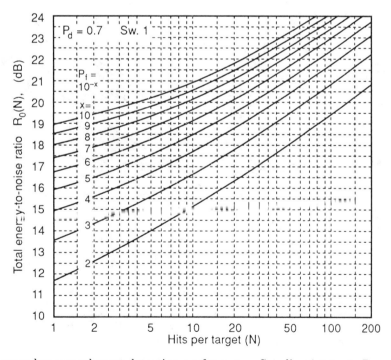

Figure 5.17: Square-law noncoherent detection performance, Swerling 1 target, $P_d = 0.75$, from [13].

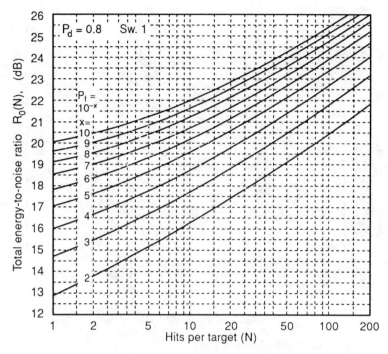

Figure 5.18: Square-law noncoherent detection performance, Swerling 1 target, $P_d = 0.8$, from [13].

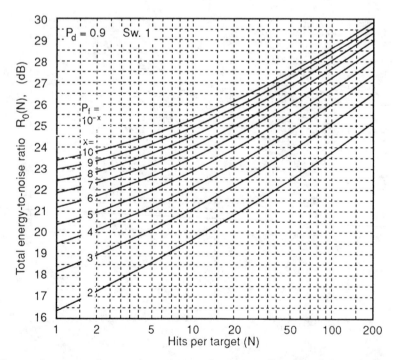

Figure 5.19: Square-law noncoherent detection performance, Swerling 1 target, $P_d = 0.9$, from [13].

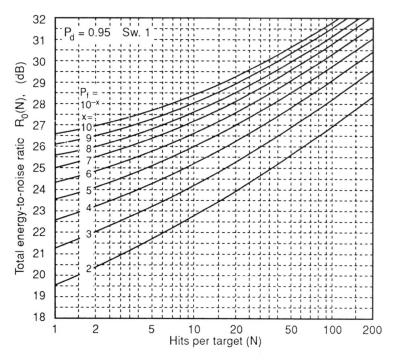

Figure 5.20: Square-law noncoherent detection performance, Swerling 1 target, $P_d = 0.95$, from [13].

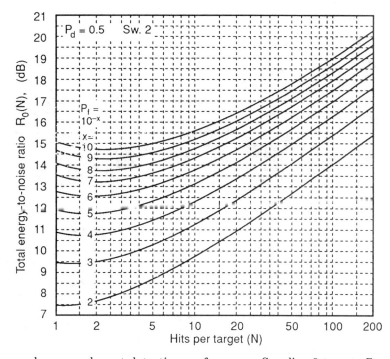

Figure 5.21: Square-law noncoherent detection performance, Swerling 2 target, $P_d = 0.5$, from [13].

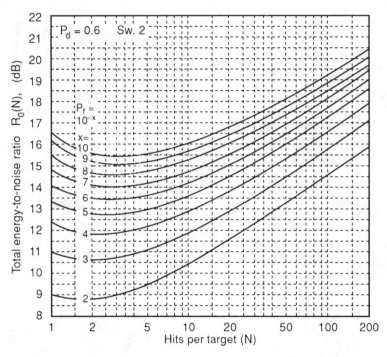

Figure 5.22: Square-law noncoherent detection performance, Swerling 2 target, $P_d = 0.6$, from [13].

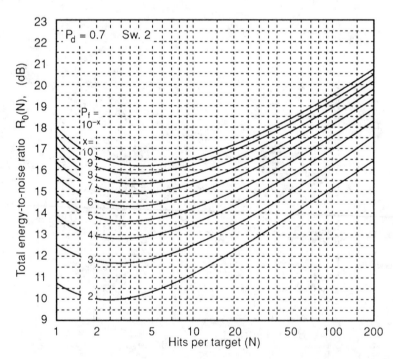

Figure 5.23: Square-law noncoherent detection performance, Swerling 2 target, $P_d = 0.7$, from [13].

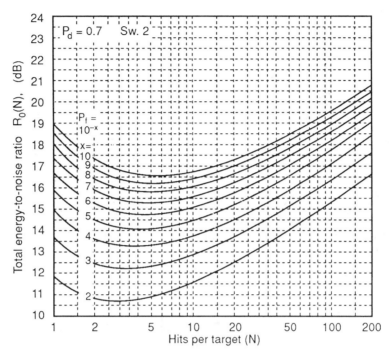

Figure 5.24: Square-law noncoherent detection performance, Swerling 2 target, $P_d = 0.75$, from [13].

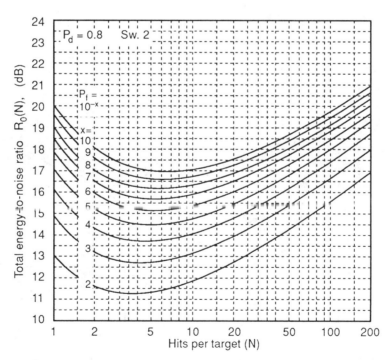

Figure 5.25: Square-law noncoherent detection performance, Swerling 2 target, $P_d = 0.8$, from [13].

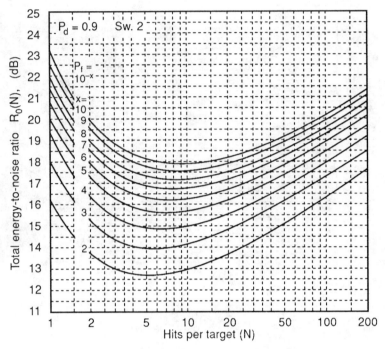

Figure 5.26: Square-law noncoherent detection performance, Swerling 2 target, $P_d = 0.9$, from [13].

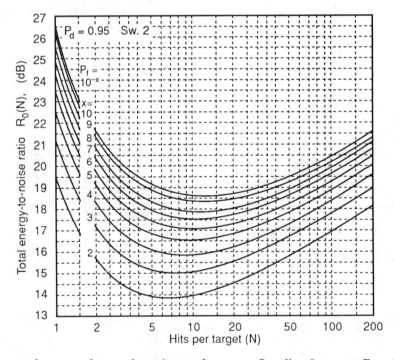

Figure 5.27: Square-law noncoherent detection performance, Swerling 2 target, $P_d = 0.95$, from [13].

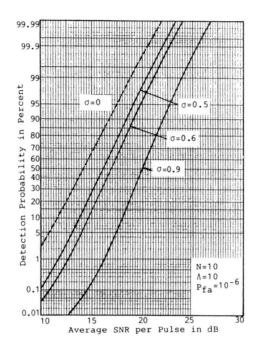

Figure 5.28: Detection performance on Swerling 2 target in exponentially correlated Rayleigh clutter, σ is correlation coefficient between adjacent samples of clutter. $N = 10$. See Table 5.4 for other conditions, from [14].

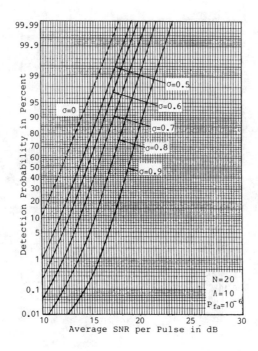

Figure 5.29: Detection performance on Swerling 2 target in exponentially correlated Rayleigh clutter, σ is correlation coefficient between adjacent samples of clutter. $N = 20$. See Table 5.4 for other conditions, from [14].

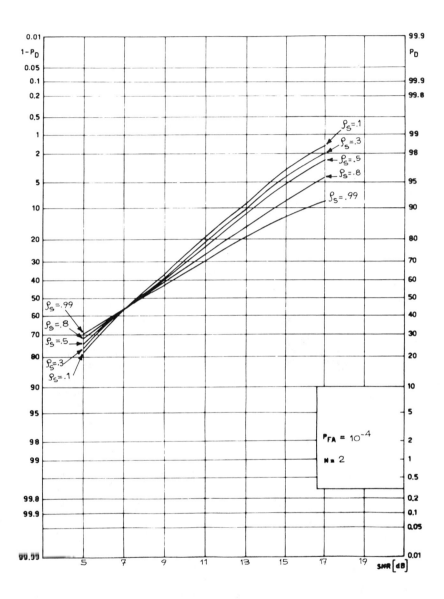

Figure 5.30: Detection performance of correlated target in Gaussian noise (target correlation coefficient ρ, as a parameter). See Table 5.4 for other conditions, from [18].

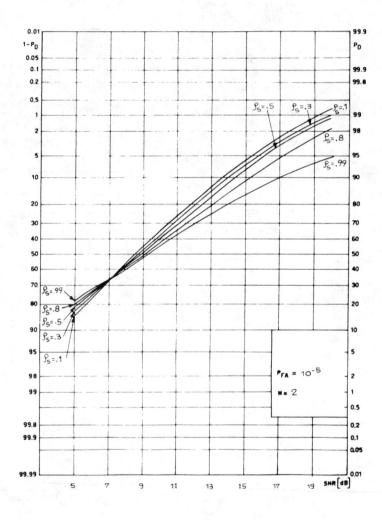

Figure 5.31: Detection performance of correlated target in Gaussian noise (target correlation coefficient ρ, as a parameter). See Table 5.4 for other conditions, from [18].

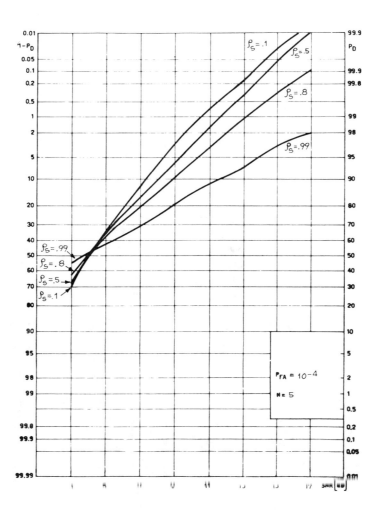

Figure 5.32: Detection performance of correlated target in Gaussian noise (target correlation coefficient ρ, as a parameter). See Table 5.4 for other conditions, from [18].

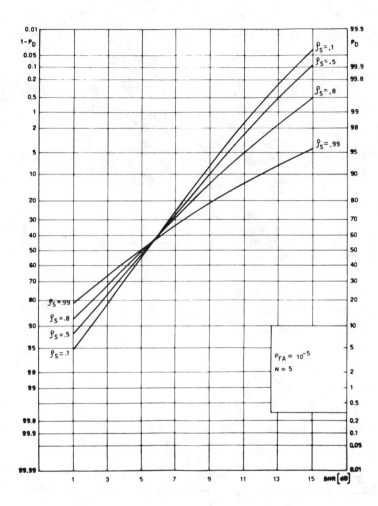

Figure 5.33: Detection performance of correlated target in Gaussian noise (target correlation coefficient ρ, as a parameter). See Table 5.4 for other conditions, from [18].

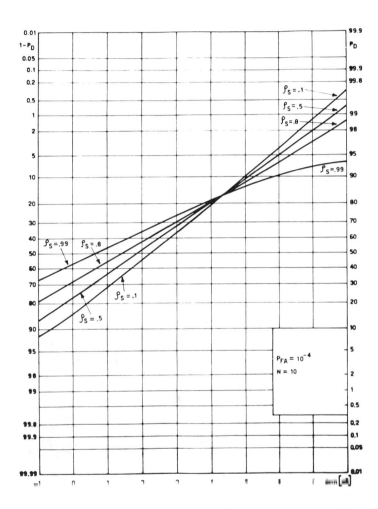

Figure 5.34: Detection performance of correlated target in Gaussian noise (target correlation coefficient ρ, as a parameter). See Table 5.4 for other conditions, from [18].

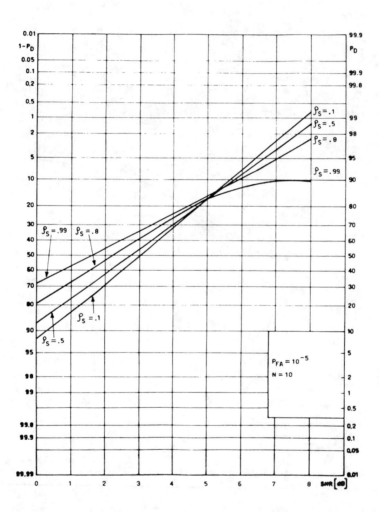

Figure 5.35: Detection performance of correlated target in Gaussian noise (target correlation coefficient ρ, as a parameter). See Table 5.4 for other conditions, from [18].

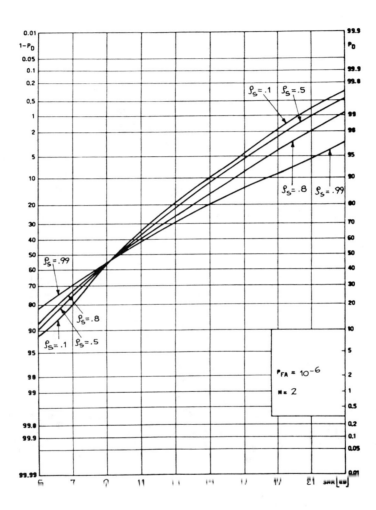

Figure 5.36: Detection performance of correlated target in Gaussian noise (target correlation coefficient ρ, as a parameter). See Table 5.4 for other conditions, from [18].

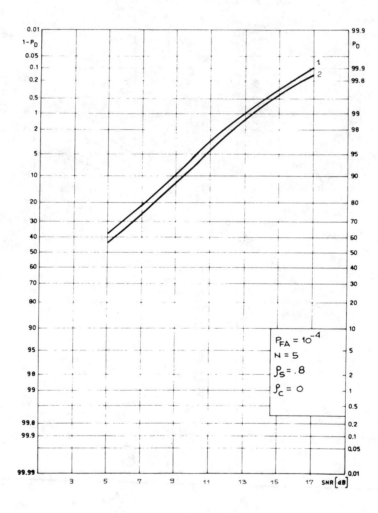

Figure 5.37: Comparison of detection performance for target with Gaussian (1) and exponential (2) shaped autocorrelation function. See Table 5.4 for other conditions, from [18].

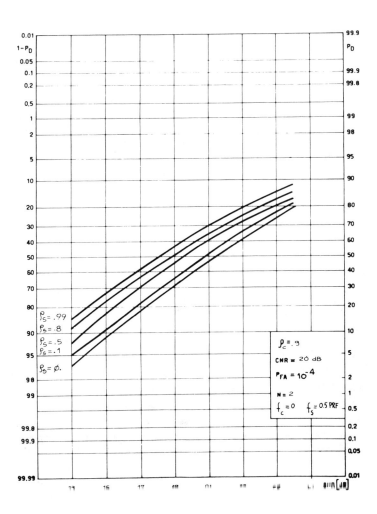

Figure 5.38: Detection performance of correlated target in clutter (target correlation coefficient ρ_s, as a parameter). See Table 5.4 for other conditions, from [18].

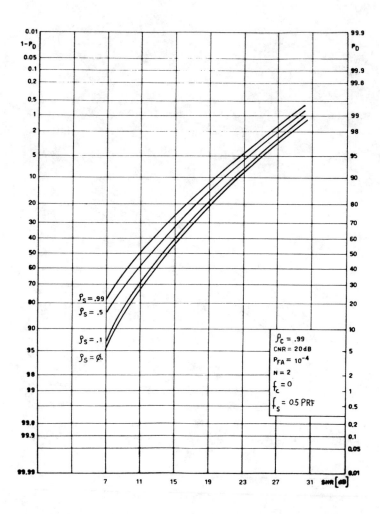

Figure 5.39: Detection performance of correlated target in clutter (target correlation coefficient ρ_s, as a parameter). See Table 5.4 for other conditions, from [18].

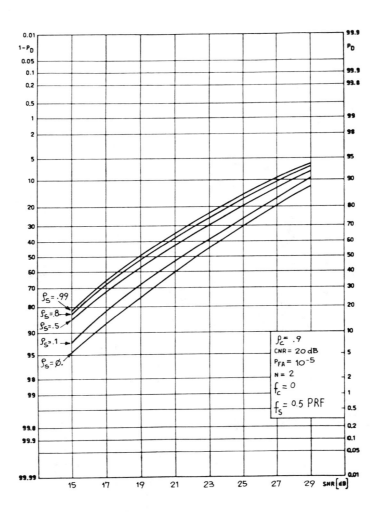

Figure 5.40: Detection performance of correlated target in clutter (target correlation coefficient ρ_s, as a parameter). See Table 5.4 for other conditions, from [18].

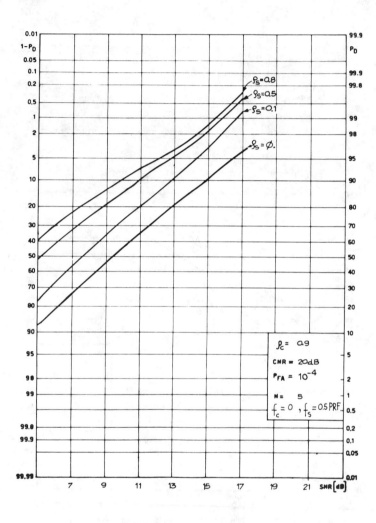

Figure 5.41: Detection performance of correlated target in clutter (target correlation coefficient ρ_s, as a parameter). See Table 5.4 for other conditions, from [18].

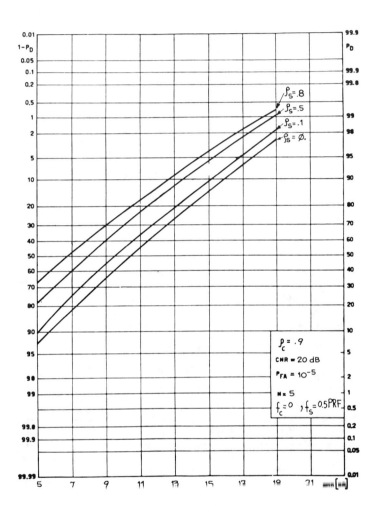

Figure 5.42: Detection performance of correlated target in clutter (target correlation coefficient ρ_s, as a parameter). See Table 5.4 for other conditions, from [18].

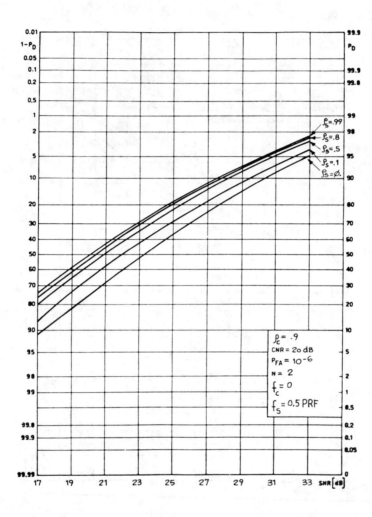

Figure 5.43: Detection performance of correlated target in clutter (target correlation coefficient ρ_s, as a parameter). See Table 5.4 for other conditions, from [18].

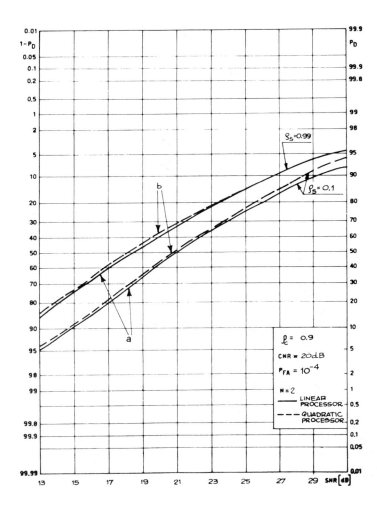

Figure 5.44: Comparison of detection performance between linear (a) and quadratic (b) processors. See Table 5.4 for other conditions, from [18].

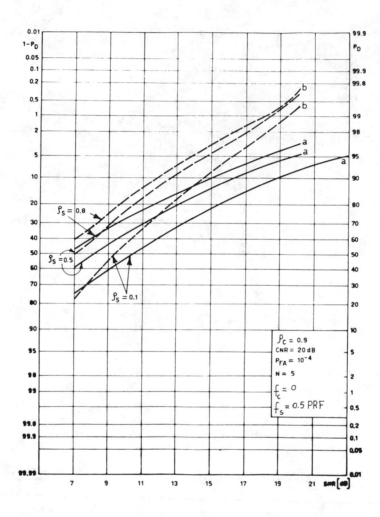

Figure 5.45: Comparison of detection performance between linear (a) and quadratic (b) processors. See Table 5.4 for other conditions, from [18].

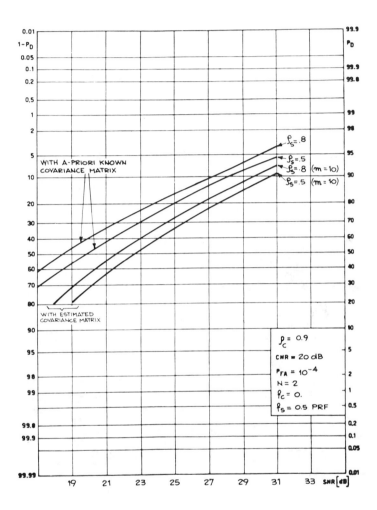

Figure 5.46: Detection loss due to adaptive estimation of disturbance covariance matrix. See Table 5.4 for other conditions, from [18].

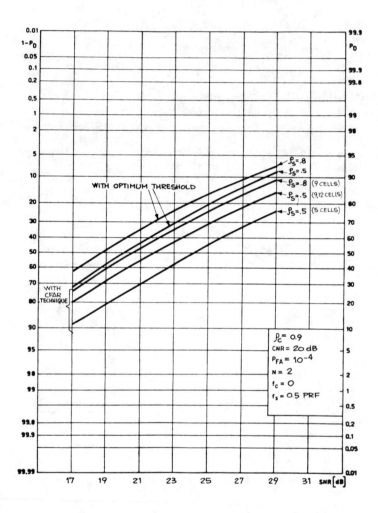

Figure 5.47: Detection loss due to CFAR thresholding. See Table 5.4 for other conditions, from [18].

5.3.2 Detection of Chi-square Targets in Clutter

Table 5.5 lists the figures that present the capability of the detection of fluctuating targets that follow the chi-square statistical distribution. The K-parameter in the chi-square distribution is the number of pairs of degrees of freedom where each degree of freedom represents an independent Gaussian distributed component in the quantity being described by the chi-square distribution. Hence, $K = 1$ corresponds to one pair of Gaussian components, giving the Rayleigh distribution.

Table 5.5: Detection probability of chi-square targets in cell-averaging CFAR.

Figure	Plot	No. of pulses N	P_{fa}	Other parameters
[15]				
Figure 5.48	P_d vs "R"	8	10^{-6}	Chi-sq.target, fixed threshold
Figure 5.49	P_d vs "R"	16	10^{-6}	Chi-sq.target, fixed threshold
Figure 5.50	P_d vs "R"	8	10^{-6}	Chi-sq.target, 8-cell CFAR
Figure 5.51	P_d vs "R"	16	10^{-6}	Chi-sq.target, 4-cell CFAR

Corresponding values are shown in Table 5.6:

Table 5.6: Values of K of the chi-square distribution which correspond with the Swerling fluctuation models.

$K=$	1	N	2	2N	∞
Swerling $=$	1	2	3	4	0 (nonfluctuating)

Table 5.7: Detection of chi-square targets with Maximum-half mean-level CFAR.

Figure	Plot	No. of pulses N	P_{fa}	Other parameters
Figure 5.52	S/N vs K parameter	5	10^{-4}	Chi-sq.target, 4,8,16-cell CFAR
Figure 5.53	S/N vs K parameter	1,2,3,5	10^{-2}	Chi-sq.target, fixed threshold
Figure 5.54	S/N vs K parameter	1,2,3,5	10^{-4}	Chi-sq.target, fixed threshold
Figure 5.55	S/N vs K parameter	1,2,3,5	10^{-6}	Chi-sq.target, fixed threshold

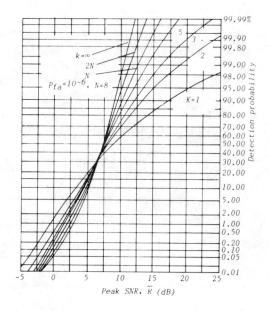

Figure 5.48: Probability of detecting fluctuating targets with a fixed threshold ($N = 8$). See Table 5.4 for other conditions, from [15].

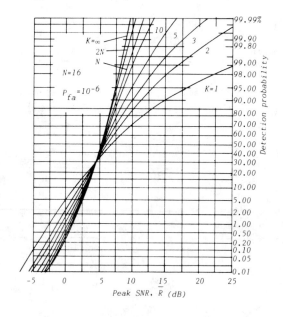

Figure 5.49: Probability of detecting fluctuating targets with a fixed threshold ($N = 16$). See Table 5.4 for other conditions, from [15].

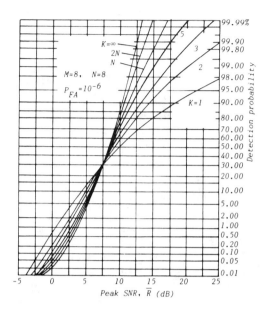

Figure 5.50: Probability of detecting fluctuating targets with a 8-cell CFAR ($N = 8$). See Table 5.4 for other conditions, from [15].

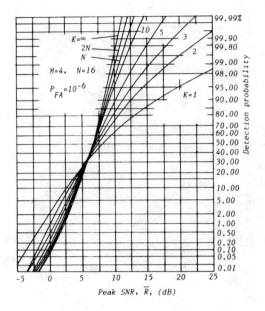

Figure 5.51: Probability of detecting fluctuating targets with a 4-cell CFAR ($N = 16$). See Table 5.4 for other conditions, from [15].

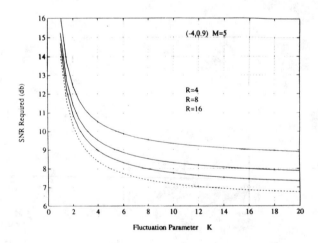

Figure 5.52: SNR required for detection of chi-square target with maximum half of mean-level CFAR using R = 4, 8, and 16-cell window. Dotted line depicts SNR required when using fixed threshold detector. From [19].

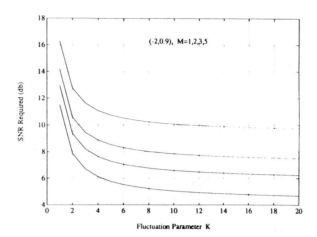

Figure 5.53: SNR required by fixed threshold detector to achieve $P_d = 0.9$ and $P_{fa} = 10^{-2}$. See Table 5.7 for conditions. From [19].

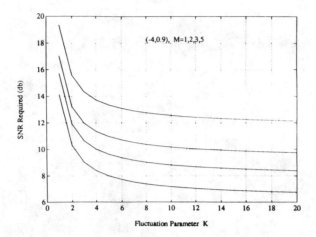

Figure 5.54: SNR required by fixed threshold detector to achieve $P_d = 0.9$ and $P_{fa} = 10^{-4}$. See Table 5.7 for conditions. From [19].

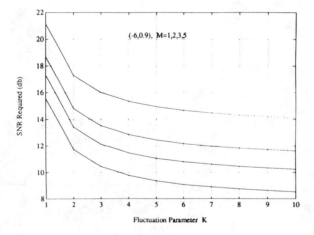

Figure 5.55: SNR required by fixed threshold detector to achieve $P_d = 0.9$ and $P_{fa} = 10^{-4}$. See Table 5.7 for conditions. From [19].

5.3.3   Detection of targets in Weibull clutter

Detection of a nonfluctuation target for various numbers of integrated pulses and Weibull parameters is shown in Figures 5.56 through 5.62. Table 5.8 defines the parameters for each of the figures.

Table 5.8: Detection of a nonfluctuating target in Weibull clutter.

Figure	Plot	No. of pulses N	P_{fa}	Other parameters
[16]				
Figure 5.56	P_d vs S/median C	1	10^{-6}	$b = .3, .4, .6, 1$, linear detector
Figure 5.57	P_d vs S/median C	3	10^{-6}	$b = .3, .4, .6, 1$, linear detector
Figure 5.58	P_d vs S/median C	10	10^{-6}	$b = .3, .4, .6, 1$, linear detector
Figure 5.59	P_d vs S/median C	30	10^{-6}	$b = .3, .4, .6, 1$, linear detector
Figure 5.60	P_d vs S/median C	3	10^{-6}	$b=.6$, linear detector
Figure 5.61	P_d vs S/median C	10	10^{-6}	$b=.6$, linear detector
Figure 5.62	P_d vs S/median C	30	10^{-6}	$b=.6$, linear detector

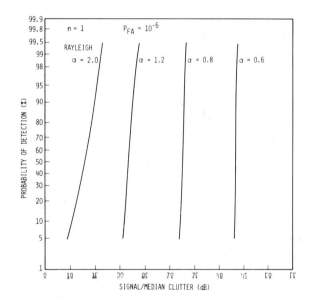

Figure 5.56: Probability of detecting nonfluctuating target with a linear detector in Weibull clutter. $\alpha = 2b$, where b is the Weibull power or RCS distribution shape parameter. Number of pulses, $N = 1$. From [16].

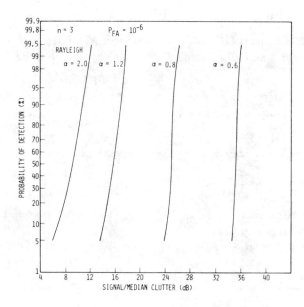

Figure 5.57: Probability of detecting nonfluctuating target with a linear detector in Weibull clutter. $\alpha = 2b$, where b is the Weibull power or RCS distribution shape parameter. Number of pulses, $N = 3$. From [16].

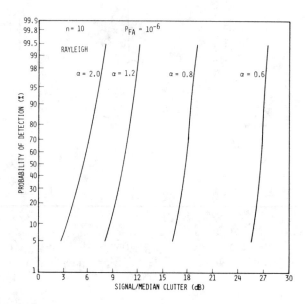

Figure 5.58: Probability of detecting nonfluctuating target with a linear detector in Weibull clutter. $\alpha = 2b$, where b is the Weibull power or RCS distribution shape parameter. Number of pulses, $N = 10$. From [16].

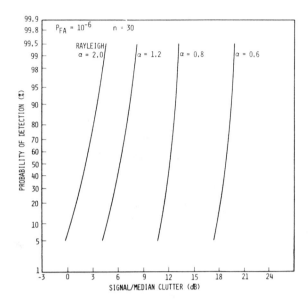

Figure 5.59: Probability of detecting nonfluctuating target in with a linear detector Weibull clutter. $\alpha = 2b$, where b is the Weibull power or RCS distribution shape parameter. Number of pulses, $N = 30$. From [16].

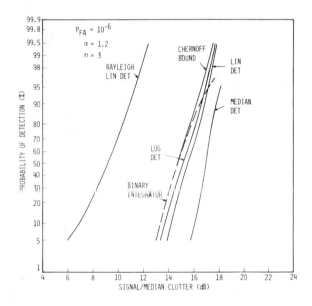

Figure 5.60: Probability of detecting a nonfluctuating target for various types of detectors in Weibull clutter with shape parameter, $b = 0.6$, and number of pulses $N = 3$. From [16].

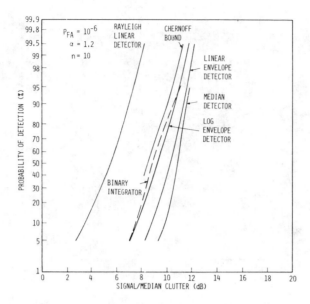

Figure 5.61: Probability of detecting a nonfluctuating target for various types of detectors in Weibull clutter with shape parameter, $b = 0.6$, and number of pulses $N = 10$. From [16].

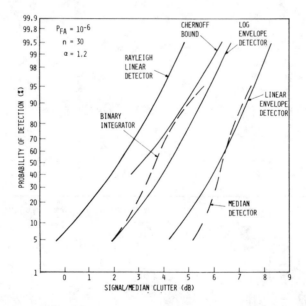

Figure 5.62: Probability of detecting a nonfluctuating target for various types of detectors in Weibull clutter with shape parameter, $b = 0.6$, and number of pulses $N = 30$. From [16].

5.3.4   Detection of Targets in k-distributed Clutter

Detection of a Swerling 2 target in k-distributed clutter is shown in Figures 5.63 through 5.66. Table 5.9 summarizes the conditions for the data. The m value shown in the figures is the shape defining parameter of the clutter distribution. It is sometimes shown as v by others. Some identifying references for m are:

1. $m = \infty$ is a Rayleigh target

2. $m = 4$ is a large reflector subject to small changes of orientation

3. $m = 0.5$ is a Weibull distribution

4. $m = 0.1$ corresponds to very long tail, spiky distribution

Table 5.9: Detection probability in k-distributed clutter.

Figure	Plot	No. of pulses N	P_{fa}	Other parameters
Figure 5.63	P_d vs S/N$_1$	10	10^{-6}	k-clutter, $m = 1$ to ∞
Figure 5.64	P_d vs S/N$_1$	20	10^{-6}	k-clutter, $m = 1$ to ∞
Figure 5.65	S/N vs k parameter	1	10^{-4}	k-distributed clutter, CNR $= -10$ dB to ∞, fixed and ideal CFAR Swerling 2 target
Figure 5.66	S/N vs k parameter	1	10^{-4}	k-distributed clutter, CNR $= -3$ dB to ∞, fixed and ideal CFAR Swerling 2 target

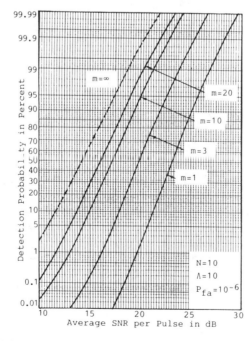

Figure 5.63: Probability of detecting a Swerling 2 target in k-distributed clutter for various shape parameter values. Number of pulses $N = 10$. From [14].

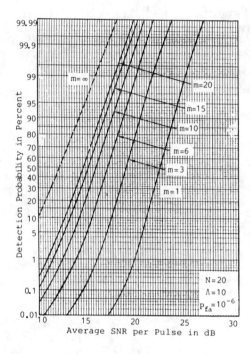

Figure 5.64: Probability of detecting a Swerling 2 target in k-distributed clutter for various shape parameter values number of pulses $N = 20$. From [14].

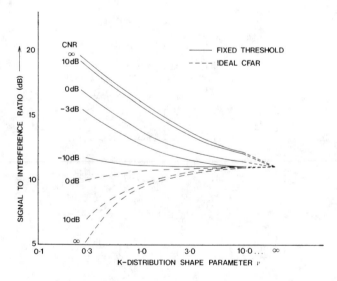

Figure 5.65: Single pulse detection of a Swerling 2 target in k-distributed clutter with $P_d = 0.5$ and $P_{fa} = 10^{-4}$. From [17].

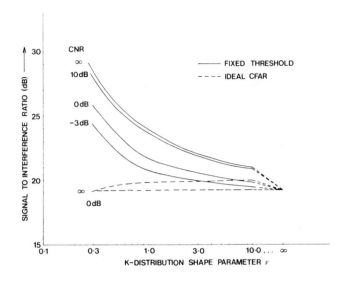

Figure 5.66: Single pulse detection of a Swerling 2 target in k-distributed clutter with $P_d = 0.9$ and $P_{fa} = 10^4$. From [17].

Bibliography

[1] Helstrom, C.W., "Approximate evaluation of detection probabilities in radar and optical communications," IEEE Trans. AES-14, no.4, July 1978, pp. 630-640.

[2] Barton, D.K., *Modern Radar System Analysis*, Artech House, Norwood, MA., 1988.

[3] Urkowitz, H., "Hansen's method applied to the inversion of the incomplete gamma function, with applications," IEEE Trans. AES-21, no. 5, Sept. 1985, pp. 728-731.

[4] Finn, H.M., "A CFAR design concept for operation in heterogenous clutter," AGARD Conf. Proc. no. 381, Multifunction Radar for Airborne Applications, July 1986, AD-A173 978, AGARD-CP-381, pp. 12-1 to 12-12.

[5] Kanter, I., "Exact detectioin probability for partially correlated Rayleigh targets," IEEE Trans. AES-22, no. 2, March 1986, pp. 184-195.

[6] Morchin, W.C., *Airborne Early Warning Radar*, Artech House, Norwood, MA, 1990.

[7] Jacobson, H.I., "A simple formula for radar detection probability on Swerling I targets," IEEE Trans. AES-17, no. 2, Mar. 1981, p. 304.

[8] Tufts, D.W., A.J. Cann, "On Albersheim's detection equation," IEEE Trans. AES-19, no. 4, July 1983, pp. 643-645.

[9] Morchin, W.C., "Airborne Early Warning Radar Studies for the V22," Final Report, Science Applications International Corp., Rep. WCM-89-8, Dec. 29, 1989, Part 2, pp. 11-13.

[10] Hardin, R.H., "System optimization techniques for radar surveillance from space," IEEE Conf. Proc. National Radar Conference, March 29 30, 1989, Dallas, TX, pp 7-12.

[11] Jacobson, H.I., "A closed form representation of the cumulative detection probability for search radars," IEEE Trans., AES-14, July 1978, pp. 699–700.

[12] DiFranco, J.V., W.L. Rubin, *Radar Detection*, Artech House, Norwood, MA, 1980.

[13] Hansen, V.G., "Topics in radar signal processing," The Microwave Journal, March 1984, pp. 24-46, and personal communication June 1984.

[14] Hou, X-Y, N. Morinaga, "Detection performance in k-distributed and correlated Rayleigh clutters," IEEE Trans. AES-25, no. 5, September 1989, pp. 634-641.

[15] Hou, X-Y, N. Morinaga, T. Namekawa, "Direct evaluation of radar detection probabilities," IEEE Trans. AES-23, no. 4, July, 1987, pp. 418-423.

[16] Schleher, D.C., "Radar detection in Weibull clutter," IEEE Trans. AES-12, no. 6, November 1976, pp. 736-743.

[17] Watts, S., "Radar detection prediction in k-distributed sea clutter and thermal noise," IEEE Trans. AES-23, no. 1, January 1987, pp. 40-45.

[18] Farina, A., A. Russo, "Radar detection of correlated targets in clutter," IEEE Trans. AES-22, no. 5, September 1986, pp. 513-532.

[19] Ritcey, J.A., "Detection analysis of the MX-MLD with noncoherent integration," IEEE Trans. AES-26, no. 3, May 1990, pp. 569-576.

Notes

Notes

Chapter 6

Pulse Doppler Systems

6.1 Introduction

This chapter is necessarily short because we are not attempting to be complete by including material that is available in existing books. Some of the more recent books provide a large amount of information, [4], [5], [6], [10], [8], [7]. In addition much that can be attributed to pulse-Doppler is contained in the chapters on MTI, Signal processing, and the radar equation. The material we give here is what has attracted our attention as useful and not widely available in book form. As we stated in the preface, future editions will include what readers may contribute and what the author adds.

6.2 Resolving Dimensions

6.2.1 Chinese Remainder Theorem for Unambiquous Solution of a Range Value

For the methods involved in solving for unambiguous range refer to [4], and [3].

6.2.2 Minimum Detectable Velocity

To obtain an estimate of the minimum detectable target velocity in clutter, we can consider the proportionality:

$$\frac{2C/N_f}{\lambda \sigma_f} \propto \frac{C/N_{mb}}{2\sigma_V} \tag{6.1}$$

for Gaussian filters and Gaussian shaped mainbeam clutter. σ_f is the rms bandwidth of a Gaussian filter, C/N_f is the clutter to noise ratio in a Doppler filter, C/N_{mb} is the mainbeam clutter to noise ratio, and σ_V is the clutter velocity spread. The term $\lambda/2$ is used to convert the Doppler filter bandwidth to an equivalent velocity bandwidth. If we consider filter attenuation of C/N_f to an acceptable level at some minimum detectable velocity, $V_{\min}$, separated from zero:

$$\frac{2C/N_f}{\lambda \sigma_f} = \frac{C/N_{mb}}{2\sigma_V} \exp\left(\frac{-V_{\min}^2}{2\sigma_V^2}\right) \tag{6.2}$$

Solving for $V_{\min}$:

$$V_{\min} = \sqrt{-2\sigma_V^2 \ln\left(\frac{2\sigma_V}{\lambda\sigma_f}\right)\left(\frac{C/N_f}{C/N_{mb}}\right)} \tag{6.3}$$

We can select $C/N_f \equiv 1$ to obtain the desired target detection at $V_{\min}$. $C/N_f = S/N$ is used in [7] to define $V_{\min}$. The expression given by [7] eq.(8.5) in which $C/N_f = S/N$, for a search airborne radar is:

$$V_{\min} = \frac{A_e\Omega}{4T_s\lambda} + 1.2\sqrt{(\frac{.42V_a\lambda}{W_A})^2 + (\frac{.21A_e\Omega}{T_s\lambda})^2} \tag{6.4}$$

where:

1. A_e is the effective antenna aperture

2. Ω is the angular search volume

3. T_s is the time to search the angular search volume

4. V_a is the radar platform speed

5. W_A is the antenna aperture width

6.2.3   Resolution of Acceleration

To resolve a range acceleration difference of $\Delta\ddot{R}$, the coherent signal must be at least as long as:

$$T_{ob} = \sqrt{\lambda/\Delta\ddot{R}}, sec \tag{6.5}$$

where T_{ob} is the coherent observation time.

The 3-dB width of the main acceleration response lobe as determined empirically by Mitchell [9] is:

$$\Delta\gamma = 3.6k_w/T_{ob}^2, Hz/sec \tag{6.6}$$

where k_w is a factor that depends upon the pulse weighting. For uniform weighting $k_w = 1$, for Hamming; $k_w \approx 1.5$.

6.2.4   Doppler Beam Sharpening

Doppler beam sharpening is often considered a class of synthetic aperture radar. However, we consider it a part of Doppler radar because it is usually a mode of a Doppler radar and not a synthetic aperture radar and because the processing does not entail the usual phase correction methods used in synthetic aperture radar. Doppler beam sharpening produces a map similar to that produced by a real-beam radar. Doppler beam sharpening, however, produces such a map at a range increased by a Doppler beam sharpening factor, [8]. We can determine the beam sharpening factor by considering the effective beam width in a Doppler beam sharpened system with the real aperture beamwidth. The effective beamwidth in a Doppler beam sharpened system is created by the angular width of the Doppler filters used. This beamwidth, [7], p. 234, as a function of antenna azimuth angle, θ_{az}, relative to the forward velocity vector of the radar platform is:

$$\beta = \lambda/2V_a \sin\theta_{az} \tag{6.7}$$

where V_a is the platform velocity. If we take the real-beam beamwidth to be λ/W_a, where W_a is the antenna width, the beam sharpening ratio is:

$$S_r = 2V_aT_{ob}\sin\theta_{az}/W_a \tag{6.8}$$

where T_{ob} is the Doppler filtering coherent processing time.

6.3 Ghosts and Number of Operations

The target returns when using a normal high PRF pulse Doppler radar are detected following first range gating then Doppler filtering. Range ambiguities are removed from these detections using some form of PRF ranging, [1], [2], [3]. However, ghosts may be created with the ranging techniques. The ghosts are false target reports created by incomplete ambiguity removal due to an interaction of noise false alarms with target detections and an interaction between the returns of multiple targets. The use of extra PRF sequences and post detection processing can significantly reduce the number of ghosts. Gerlach and Andrews, [3] studied a particular method for reducing ghosts using three stages of thresholding prior to using a deghoster processor. The first threshold for detection was applied to the output of each range-gate Doppler filter combination. The second threshold for detection was applied as requiring M detections out of N PRFs at the output of a range ambiguity resolver and the third to a square law detector that integrated the contribution of each PRF for each unambiguous range. A candidate target was declared if all thesholds were exceeded. Their deghosting scheme processed the candidate targets. They found that the number of software operations and ghosts varied inversely with the SNR required for target detection. Their simulation results are shown in Figures 6.1, 6.2 and 6.3. 100 targets were assumed to be in the antenna mainbeam and distributed in 100 Doppler bins and 1200 range bins in the simulation. The results are shown plotted as a function of the probability of false alarm at the output of the M out of N detector. The P_D and P_F are respectively the probability of detection and false alarm at the output of the square law detector.

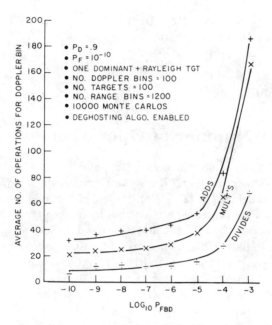

Figure 6.1: Number of software operations required per Doppler filter as determined by Monte Carlo simulation, from [3].

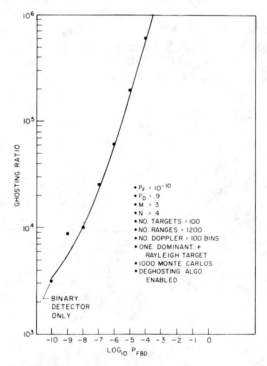

Figure 6.2: Ghosting ratio (ghosting probability to quiescent noise only false alarm probability) as determined by Monte Carlo simulation, from [3].

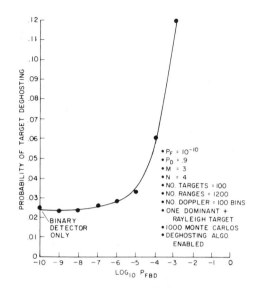

Figure 6.3: Target dehosting (probability that real target will be declared as a ghost) as determined by Monte Carlo simulation, from [3].

Bibliography

[1] Skillman, W.A., "Pulse Doppler Radar," Chap. 17 of *Radar Handbook*, M. Skolnik, ed., McGraw-Hill, New York, NY, 1990.

[2] Hovanessian, S.A., "An algorithm for calculation of range in a multiple PRF radar," IEEE Trans. AES- 12, no. 2, March 1976, pp. 287-290.

[3] Gerlach, K., G.A. Andrews, "Cascaded detector for multiple high-PRF pulse Doppler radars," IEEE Trans. AES-26, no. 5, September 1990, pp. 754-767.

[4] Long, W.H., D.H. Mooney, W.A. Skillman, "Pulse Doppler radar," Chap.17 of *Radar Handbook*, M. Skolnik, ed., McGraw-Hill, 1990.

[5] Mooney, D.H., W.A. Skillman, "Pulse Doppler radar," Chap. 19 of *Radar Handbook*, M. Skolnik, ed., McGraw-Hill, 1970.

[6] Stimson, G.W., *Introduction to Airborne Radar*, Hughes Aircraft Company, 1983.

[7] Morchin, W.C., *Airborne Early Warning Radar*, Artech House, Norwood, MA, 1990.

[8] Morris, G.V., *Airborne Pulsed Doppler Radar*, Artech House, Norwood, MA, 1988.

[9] Mitchell, R.L., "Resolution in Doppler and acceleration with coherent pulse trains," IEEE Trans. AES- 7, no. 4, July 1971, pp. 630-636.

[10] Nathanson, F.E., *Radar Design Principles*, McGraw-Hill, New York, NY, 1969.

Notes

Notes

Chapter 7

Electronic Countermeasures

7.1 Introduction

Electronic countermeasures (ECM) is a division of electronic warfare that involves actions taken to prevent or reduce an enemy's effective use of the electromagnetic spectrum, [2]. Conversely, electronic counter-counter measures (ECCM) involves actions taken to insure friendly effective use of the eletromagnetic spectrum despite the enemy's use of electronic warfare. A third division is ESM - electronic warfare actions involving search, intercept, identification and location of sources of radiated electromagnetic energy for immediate threat recognition. We include topics associated with each of the three divisions in this chapter under a broader definition of electronic countermeasures used for the chapter title.

7.2 Equations

7.2.1 Target Signal to Jammer Noise ratio

In terms of radar average power-aperture product, the target signal (S) to jamming signal (J) ratio, $J >>$ thermal noise, is:

$$\frac{S}{J} = \frac{P_{avg} A_e \sigma}{P_j G_j g_{rs} df \lambda^2 L_r L_j} \frac{R_j^2}{R_t^4} \frac{B_j}{B_{IF}} \tag{7.1}$$

where

P_{avg} is the radar average power

A_e is the radar transmit antenna area

σ is the target radar cross section

P_j is the jammer rf power

G_j is the jammer antenna gain in the direction of the radar

g_{rs} is the receive sidelobe antenna gain of the radar relative to the mainbeam gain in the direction of the jammer. $g_{rs} = 1$ when the jammer is in the mainbeam.

df is the radar duty factor

L_r is the radar system loss to target signal

L_j is the loss from the jammer power point to the radar IF

R_j is the range between the jammer and the radar

R_t is the range between the radar and the target

B_{IF} is the radar IF bandwidth

B_j is the jammer rf bandwidth

The reader may use the search and track radar range equations in Chapter 16, Radar Equation, to solve for the power-aperture product. Additionally, the reader is referred to Farina, [1] section 9.12, for specific search or track equations for determining detection range in the presence of noise jamming.

7.2.2   ECCM Improvement Factors

Improvement factors are measures of relative performance of particular ECCM methods and not indicators of actual performance, [2]. The factors are most applicable to power related ECM methods and are useful for comparing design changes. The equations dealing with integration assume noncoherent integration as formulated. Should the application use coherent integration the user should use $10\log(N_2/N_1)$ rather than $5\log(N_2/N_1)$, where N_2/N_1 are as defined below.

Frequency agility $EIF_{FA} = 10\log\Delta F - 10\log B_{rf}$ with the assumption that the jammer can spot jam B_{rf} in the absence of frequency agility, and must jam ΔF when agility is used.

   ΔF is the frequency agile band

   τ is the radar pulsewidth

   B_{rf} is the radar receiver rf bandwidth

Peak power $EIF_{PK} = 10\log P_2/P_1$

   P_i is the peak power where $i = 2$ indicates the value after a design change and $i = 1$ indicates the value prior to a design change.

PRF $EIF_{PRF} = 10\log PRF_2/PRF_1 + 5\log N_2/N_1$

   PRF_i is the PRF, N_2/N_1 is the ratio of the number of noncoherent pulses or noncoherent pulse intervals integrated.

Average power output $EIF_{AVG} = EIF_{PK} + EIF_{PRF} + 10\log \tau_2/\tau_1$

Pulsewidth for noise jamming $EIF_{\tau J} = 10\log(\tau_2/\tau_1) - 10\log(k_{f1}/k_{f2})$

   k_{fi} is a filter matching factor, [12], or likewise an efficiency [13], or the inverse of filter matching loss, [14]. We note that this and the next formula correct those (equations 9.6 and 9.7) in [2].

Pulsewidth for chaff $EIF_{\tau C} = 10\log(\tau_1/\tau_2) - 10\log(k_{f2}/k_{f1})$

Pulse compression $EIF_{PC} = 10\log C_r$

    C_r is the pulse compression ratio.

Pseudonoise $EIF_{PN} = 20\log(2^n - 1)$

    n is the number of shift register stages used to generate the pseudo noise.

Sidelobe canceller $EIF_{SLC} = 10\log CR$

    CR is the sidelobe cancellation ratio.

Low-sidelobe antenna in noise jamming $EIF_{sllJ} = sll_2 - sll_1 - 2(G_1 - G_2) + 5\log N_2/N_1$ sll_i is the receive sidelobe level relative to the mainbeam gain, dB. G_i is the mainbeam gain.

Low-sidelobe antenna for the jamming free (benign) condition $EIF_{sllB} = sll_2 - sll_1 - 2(G_1 - G_2) + 5\log N_2/N_1$

Low-sidelobe antenna for chaff $EIF_{sllC} = 10\log\theta_1/\theta_2 + 5\log N_2/N_1$

    θ_i is the azimuth beamwidth. This assumes the chaff corridor is much smaller in height than the height covered by the elevation beam.

MTI against chaff $EIF_{MTI} = I_{dB}$

    I_{dB} is the MTI improvement factor. It may be better to determine I_{dB} for the restricted range of possible chaff velocities when a high-PRF design is being evaluated rather than the normal practice of using the range of target velocities. EIF_{MTI} in high-PRF applications may not show a significant difference between chaff and non-chaff situations. In addition, it may be worthwhile to consider subtracting a ΔI_{dB} corresponding to a degradation of I_{dB} on targets if a special chaff MTI notch is used.

7.3   Expendables

7.3.1   Chaff

The average radar cross section will be:

$$\sigma \approx 0.17\lambda^2 N \tag{7.2}$$

where N is the number of half-wave chaff dipoles.

$$N = 150 W_d F_{GHz} p_f / W_l \tag{7.3}$$

where W_d is the weight (kg) of dispensed dipoles, F_{GHz} is the dipole resonant frequency (GHz), p_f is the normalized packing density, W_l is the weight (gm) per meter of a chaff dipole. Common values are given in Table 7.1 [3].

Table 7.1: Comparison of properties of common chaff types, after [3].

Chaff type	Section dimensions of filaments (μm)	Density (kg/m³)	p_f	W_l (g/m)	Mean fall rate still air at sea level (m/s)
Aluminized Glass	25	2550	.55	0.010	0.3
Silverized nylon monofilament	90	1300	.65	0.066	0.6
2 × 1 aluminum foil	50 × 25	2700	.55	0.0068	.40-.45
4 × 1 aluminum foil	100 × 25	2700	.55	0.014	.40-.45
8 × 1/2 aluminum foil	200 × 12	2700	.45	0.013	.40-.45

A chaff dipole will provide minor peak responses at multiples of a half-wavelength. When dipoles of different lengths are mixed the total RCS will be an addition of the contributions from the different lengths. Figure 7.1 shows an example of the individual and combined RCS values.

Spectral characteristics of aircraft dispensed chaff

Estes et al. [4] present measurements of the range rate and velocity spread of jet aircraft dispensed chaff as viewed from a ground radar. The aspect angle of the measuremnets were 8 to 20° relative to the aircraft heading. Their data are summarized in Table 7.2. They concluded the average inertial space radial velocity, $\dot{R}$, of X-band chaff nearly reached the ambient wind velocity of within about 200 m behind the aircraft. They found that it took longer for UHF chaff to reach the ambient wind velocity. These data are illustrated by the values in the first column of Table 7.2. One should only consider the change in $\dot{R}$ values because absolute values depend upon ambient winds at the time of measurement.

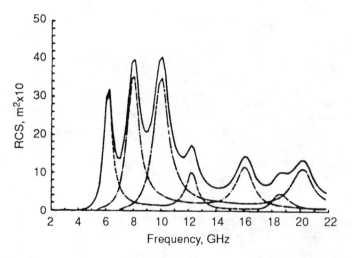

Figure 7.1: Theoretical response curves for dipoles resonant at 6, 8, and 10 GHz. Number at each respective frequency is: 0.8-, 1.5-, and 2.3 million. From [3].

Table 7.2: Chaff range rate and velocity spread in the wake of an approaching jet aircraft, after [4]. Radar line-of-sight is within aspect angles of 8 to 20° of aircraft heading. Aircraft range rate towards radar is about 110 m/s.

$\dot{R}$ (m/s)	σ_v (m/s)	Distance behind aircraft (m)	Chaff frequency (band)
8.1	1.2	210	X
8.1	1.1	420	
9.9	0.6	14km	
1.3	2.0	150	UHF
2.0	1.5	300	
2.8	1.5	450	
4.1	1.2	6.2 km	

7.3.2   Decoys

The fractional part of seeker sum channel one-way beamwidth in the presence of a Swerling 1 target and a constant amplitude repeater towed decoy is, [5] and [6]:

$$\bar{r} = \frac{r_1 + r_2}{2} - \frac{r_2 - r_1}{2} + \frac{1}{N} \sum_{i=1}^{N} \frac{1 - K^2}{1 + 2K \cos \phi_i + K^2} \tag{7.4}$$

where:

r_1 **and** r_2 are the angular positions of the target and decoy in terms of the fractional parts of the seekers one-way half- power beamwidth

K is the voltage ratio of the target and decoy as modified by the seeker antenna gain

N is the number of independent samples processed by the seeker

ϕ is the phase shift between the target and decoy at the seeker

The summation is replaced by the equivalent integral form for a CW application. When $K < 1, / \bar{r} = r_1$. When $K > 1$ $\bar{r} = r_2$ and [5]:

$$K = \frac{4\pi ERP_J G_s^2(\phi_s - \phi_d) R_R^2}{ERP_R \sigma_t G_s^2(\phi_s - \phi_a)} \tag{7.5}$$

where

ERP_J is the decoy repeater effective radiated power

ERP_R is the effective radiated power of the illuminator radar for a semi-active missile seeker

G_s is the missile seeker antenna one-way power gain

$\phi_s - \phi_d$ is the angular difference between the line-of-sights of the missile seeker antenna boresight and the decoy.

$\phi_s - \phi_a$ is the angular difference between the line-of-sights of the missile seeker antenna boresight and the target.

R_R is the range between the target and the illuminating radar for the missile seeker

σ_t is the bistatic radar cross section of the target for the illuminating radar to missile seeker situation

In the active missile situation, K can be determined by replacing R_R with the range between the target and the missile seeker and using the target monostatic RCS for σ_t.

The reader is referred to [7] for some example applications that determine the required decoy tow line length and power for specified missile miss distances.

The decoy repeater gain required to emulate a particular σ_t is, [8]:

$$G_d = \frac{4\pi\sigma_t}{\lambda^2} \qquad (7.6)$$

7.4   Sidelobe Cancellation

The added loss to target signal due to overmatching the sidelobe canceller degrees of freedom is, [9]:

$$L_{slc} = \frac{K - L + 2}{K - N + 2} \qquad (7.7)$$

where N is one more than the number of auxiliary channels, and K is the number of input samples used by each auxiliary to sample noise signals, $K \leq N - 1$. L is the order of the canceller for which $L < N$ and where the noise residual in the main channel increases for order $> L$.

7.5   ESM - Electronic Warfare Support Measures

7.5.1   LPI-Low Probability of Intercept

The intercept receiver probability of detection, $Q_D < 0.1$, of a radar signal in a single cell of resolution is, after [10]:

$$
\begin{aligned}
Q_D &\approx \exp[7.33 + 1.07\ln Q_F - (0.193 + 0.064\ln Q_F)r] \\
r &\approx [1 + (1 + 9.2U)^2]/2U \\
U &= (S + 2.3TB)/S^2, \\
S &= S_r/N_i
\end{aligned}
\qquad (7.8)
$$

where S_r is the radar signal at the intercept receiver and N_i is the noise in the intercept receiver bandwidth.

Q_F is the probability of false alarm at the intercept receiver and T is the time duration that must be spent searching for a radar signal and B is the frequency band spread that must be searched. The approximation for low values of Q_D are appropriate because the radar users objective is to not be detected.

$$S_r = \frac{P_k G g_{sl} A_i}{4\pi R_i^2 L_i} \tag{7.9}$$

where P_k is the radar peak power, G is the radar antenna gain, g_{sl} is the radar sidelobe level relative to the mainbeam gain, A_i is the intercept system antenna aperture area, R_i is the range from the radar to the intercept system, and L_i are the system losses from the radar to the intercept detector.

The cumulative probability of an intercept over a period of time is:

$$P_{ci} = 1 - (1 - Q_D)^{m_i} \tag{7.10}$$

where m_i is the total number of resolution cells containing the radar signal. $m_i = T_d B / T_r B_r$, $T_r B_r$ is the radar time-bandwidth product and T_d is the intercept decision duration and B the band spread the intercept must search.

7.5.2 Instantaneous Frequency Measurement (IFM) Receiver Design

The necessary gain preceeding the IF logarithmic amplifier is, [11]:

$$G = NP/FkT_oB_r \tag{7.11}$$

where NP is the noise power at the input to logarithmic amplifier, F is the system noise figure (numeric) ahead of the logarithmic amplifier, T_o is $290^o K$, k is Boltzmanns constant (1.38×10^{-23}) and B_r is the rf bandwidth.

The tangential sensitivity is:

$$T_s = kF[6.31B_V + 2.5\sqrt{(2B_r B_V - B_V^2)}] \tag{7.12}$$

where B_V is the video bandwidth.

The frequency resolution is:

$$\delta f = (2^n t_d)^{-1} \tag{7.13}$$

where n is the number of bits used for the fine-phase discriminator, and t_d is the delay time of the delay line that provides phase differential to the phase detector.

The frequency error is:

$$\epsilon_f \approx \delta f / \sqrt{12} \tag{7.14}$$

7.6   ECM, ECCM, and ESM equipment

System related weight, volume and prime power interrelationships of airborne EW equipment are shown in Figure 7.2 and Figure 7.3.

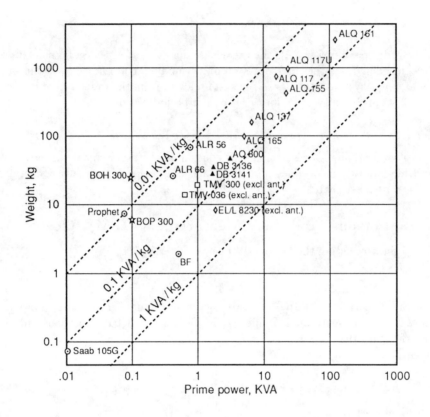

Figure 7.2: Weight and prime power relationship of airborne EW equipment; ○ refers to warning receivers, □ refers to ESM equipment, △ refers to noise jammers, ⋆ refers to chaff dispensers with chaff, and ◇ refers to deception jammers. Open symbols are for internally mounted equipment and solid symbols are for pod mounted equipment.

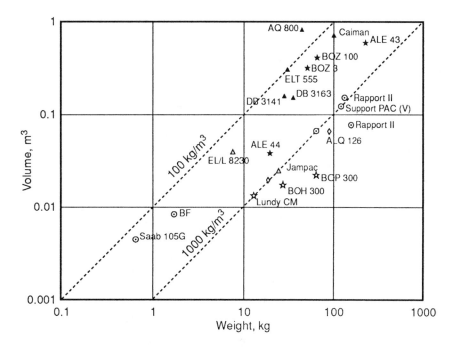

Figure 7.3: Volume and Weight relationship of airborne EW equipment; o refers to warning receivers, □ refers to ESM equipment, △ refers to noise jammers, ⋆ refers to chaff dispensers with chaff, and ◇ refers to deception jammers. Open symbols are for internally mounted equipment and solid symbols are for pod mounted equipment.

Bibliography

[1] Farina, A., "Electronic Counter Measures" *Radar Handbook*, M. Skolnik, ed., McGraw-Hill, New York, 1990.

[2] Johnstone, S.L., "ECCM for AEW Radar," Ch. 9 *Airborne Early Warning Radar*, W. Morchin, ed., Artech House, Norwood, MA, 1990.

[3] Butters, B.C.F., "Chaff," IEE Proc. Vol. 129, Pt. F, no. 3, June 1983, pp. 197-201.

[4] Estes, W.J., R.H. Flake, C.C. Pinson, "Spectral characteristics of radar echoes from aircraft dispensed chaff," IEEE Trans. AES-21, no. 7, January 1985, pp. 8-20.

[5] Kerins, W.J., "Towed decoys - will they protect against a variety of radar threats?," Journal of Electronic Defense, April 1990, pp. 59-63.

[6] Kanter, I., "Varieties of average monopulse responses to multiple targets," IEEE Trans. AES-17, no. 1, January 1981, pp. 25-28.

[7] Menegozzi, L., W.High, "Towed-decoy effectiveness: A kinematic model," Journal of Electronic Defense, April 1990, pp. 64-74.

[8] Meadows, L.A., "Expendable repeater decoy isolation techniques," Journal of Electronic Defense, December 1990, pp. 47-50.

[9] Gerlach, K., F.F. Kretshmer, "Convergence properties of Gram-Schmidt and SMI adaptive algorithms," IEEE Trans. AES-26, no. 1, January 1990, pp. 44-56.

[10] Dillard, R.A., Dillard, G.M., *Detectability of Spread-Spectrum Signals*, Artech House, Norwood, MA, 1989.

[11] Williams, R., "Theory and application of a 0.5 to 18 GHz tunable IFM receiver," Microwave Journal, February 1989, pp. 89-108.

[12] Brookner, E., *Radar Technology*, Artech House, Norwood, MA, 1980, p. 407.

[13] Skolnik, M.I., *Introduction to Radar Systems*, McGraw-Hill, New York, 1980, p. 374.

[14] Blake, L.V., "Prediction of Radar Range," Ch. 2 *Radar Handbook*, M. Skolnik, ed., McGraw-Hill, New York, 1990.

Notes

Notes

Chapter 8

General

8.1 Radar and ECM Band Nomenclature

The U.S. defense department, FAA, NASA, and commercial radar use IEEE STD 521-1976 the standard letter designations for radar-frequency bands as shown in Table 8.1.

Table 8.1: Standard radar-frequency letter band nomenclature, from [2].

Band Designation	Nominal Frequency Range	Specific Frequency Ranges for Radar Based on ITU Assignments for Region 2, see Note (1)
HF	3 MHz–30 MHz	Note (2)
VHF	30 MHz–300 MHz	138 MHz–144 MHz
		216 MHz–225 MHz
UHF	300 MHz–1000 MHz (Note 3)	420 MHz–450 MHz (Note 4)
		890 MHz–942 MHz (Note 5)
L	1000 MHz–2000 MHz	1215 MHz–1400 MHz
S	2000 MHz–4000 MHz	2300 MHz–2500 MHz
		2700 MHz–3700 MHz
C	4000 MHz–8000 MHz	5250 MHz–5925 MHz
X	8000 MHz–12 000 MHz	8500 MHz–10 680 MHz
K_u	12.0 GHz–18 GHz	13.4 GHz–14.0 GHz
		15.7 GHz–17.7 GHz
K	18 GHz–27 GHz	24 05 GHz–24.25 GHz
K_a	27 GHz–40 GHz	33.4 GHz–36.0 GHz
V	40 GHz–75 GHz	59 GHz–64 GHz
W	75 GHz–110 GHz	76 GHz–81 GHz
		92 GHz–100 GHz
mm (Note 6)	110 GHz–300 GHz	126 GHz–142 GHz
		144 GHz–149 GHz
		231 GHz–235 GHz
		238 GHz–248 GHz (Note 7)

It is to be noted that the band, 216 to 225 MHz in region 2, may not be authorized beyond January 1, 1990 [1].

The standard reported, [3] for the ECM bands and channels are given in Table 8.2. Each band is divided into 10 numerical channels, each with a channel width as shown. The beginning channel of the numerical sequence starts with 1.

8.2   Millimeter Wave to Ultraviolet Wavelength Regions

The frequency allocations, Table 8.3, were guided by the defined division of the 40 to 275 GHz frequency region into atmospheric windows and absorption bands as shown in Table 8.4.

Table 8.5 shows the wavelength regions for infrared and higher frequencies, [5].

Table 8.2: ECM operational band and channel nomenclature.

Band	Frequency (MHz)	Channel width (MHz)
A	0-250	25
B	250-500	25
C	500-1000	50
D	1000-2000	100
E	2000-3000	100
F	3000-4000	100
G	4000-6000	200
H	6000-8000	200
I	8000-10,000	200
J	10,000-20,000	1000
K	20,000-40,000	2000
L	40,000-60,000	2000
M	60,000-100,000	4000

Table 8.3: Radiolocation (radar) allocations and shared services above 40 GHz.

Frequency band	Radar status	Other services
59-64	Primary	FIXED, MOBILE, INTER-SATELLITE
76-81	Primary	Amateur, amateur-satellite, satellite services, and space radar research for earth exploration
92-95	Primary	FIXED, FIXED-SATELLITE, MOBILE
95-100	Secondary	MOBILE, MOBILE-SATELLITE, RADIONAVIGATIOIN, NAVIGATION-SATELLITE
126-134	Primary	FIXED, MOBILE, INTER-SATELLITE
134-142	Secondary	MOBILE, MOBILE-SATELLITE, RADIO-NAVIGATION, RADIO-NAVIGATION-SATELLITE
144-149	Primary	Amateur, amateur-satellite
231-235	Secondary	FIXED, FIXED-SATELLITE(space to earth), MOBILE
238-241	Secondary	FIXED, FIXED-SATELLITE (space to earth), MOBILE
241-248	Primary	Amateur, amateur satellite

Capital letters indicate primary allocations.

Table 8.4: Atmospheric window and absorption band limits, from [4].

Window Band	Limits (GHz)	Absorption band	Limits (GHz)
W_1	39.5-51.4	A_1	51.4-66
W_2	66-105	A_2	105-134
W_3	134-170	A_3	170-190
W_4	190-275		

Table 8.5: Wavelengths for infrared and higher frequency bands, after [5].

Region	Wavelength (μm)
Far infrared	14-50
Longwave infrared	8-14
Middle infrared	3-8
Near infrared	1-3
Very near infrared	0.76-1
Visible	0.38-0.76
Ultraviolet	0.1 -0.38

8.3 Descriptions of Regions of the Growth Curve

A somewhat arbitrary S-shaped curve is often used to protray the stages of growth. Figure 8.1 is a representation of such a curve the author has taken note of, but for which he can not recall the reference. The stages of growth are described:

1. **Invention or concept**: Alternate methods of achieving the same means are investigated. Feasibility studies show promise but unacceptable performance. The state of related technology is frequently controlling, especially in materials.

2. **Sound idea**: One or more working designs exist. A sifting out of best approaches occurs.

3. **Consolidation and approach to maturity**: The number of competiting concepts is reduced by superior ones. Winning designs have been established.

4. **Maturity**: Product improvements are minor and relatively expensive.

8.4 Statistical Descriptors

The standard deviation $\sigma = \sqrt{m_2}$, where m_2 is the second moment. Each moment, m_i is:

$$m_i = \sum_{all\ k} p_k (x_k - \bar{x})^i \tag{8.1}$$

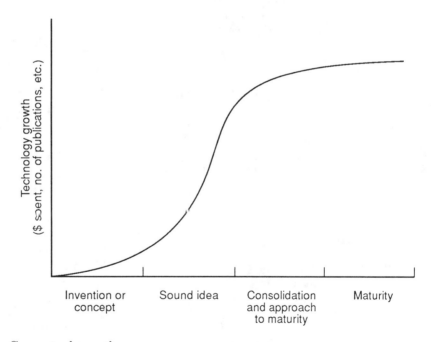

Figure 8.1: Conceptual growth curve.

where p_k is the probability of obtaining x_k as the result of one trial. For uniform probabililty, $p_k = 1/n$ for n variates, x_k. $\bar{x} = n \sum_k p_k x_k$ is the mean value.

The skewness $= m_3/\sigma^3$

1. nonzero skewness indicates lack of symmetry around the mean

2. negative skewness indicates skewness to the left

3. positive skewness indicates skewness to the right

The kurtosis $= (m_4/\sigma^4) - 3$

1. is a measure of peakedness

2. is zero if Gaussian

3. is positive if more peaked than Gaussian

4. is negative if less peaked than Gaussian

8.4.1   Estimating Rules

An estimate of the expected value is:

$$E \approx \frac{A + 4M + B}{6}$$

A is the smallest possible value
M is the most likely value
B is the largest possible value

(8.2)

An estimate of the standard deviation is:

$$\sigma \approx \frac{B - A}{6}$$

(8.3)

for large ($n > 400$) number of samples. For a limited number of samples the divisor in the above equation decreases as a function of the number of samples. An approximation, from data in [6] and [7], for the divisor ($n < 400$) is:

$$\text{Divisor} \approx 10^{\frac{\sqrt{\log n}}{2}}$$

(8.4)

8.5   The -ilities

8.5.1   Inherent Availability

The probability that the equipment will operate satisfactorily at any given time without consideration for scheduled or preventive maintenance but in an ideal support environment is [8]:

$$A = \frac{MTBF}{MTBF + MTTR}$$

(8.5)

where $MTBF$ is the mean number of life units during which all parts of the item perform within their specified limit for a specified measurement interval. $MTTR$ is the sum of the active repair times divided by the total number of malfunctions during a given period of time.

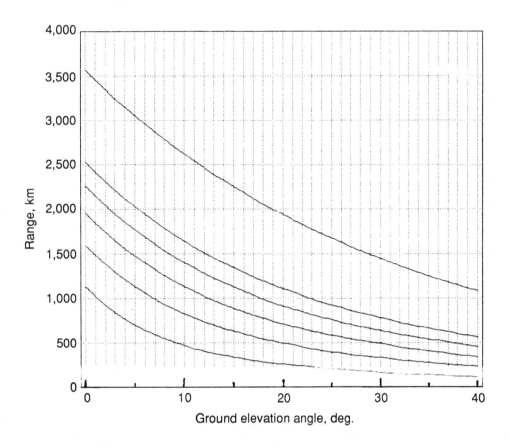

Figure 8.2: Maximum range from satellite to ground point as a function of angle above the ground local horizon. Data shown plotted for satellite altitudes starting at top: 1000-, 500-, 400-, 300-, 200-, and 100-km.

8.6 Earth and Satellite Related Subjects

8.6.1 Satellite to Ground Range, Angle, and Velocity

The maximum range between a ground point and a satellite is [11]:

$$R_s = \sqrt{r_e^2 \tan^2 \phi + 2r_e h_s} - r_e \tan \phi \tag{8.6}$$

where r_e is the radius of the earth ($r_e \approx 6371$ km), h_s is the satellite height above the earth, and ϕ is the angle above the local horizon at the ground point. An example solution is shown in Figure 8.2.

The satellite velocity for circular orbits is [9]:

$$V_s = \frac{V_e}{\sqrt{1 + h_s/r_e}} \tag{8.7}$$

where $V_e = 7.913 km/s$.

Geostationary altitude is 35,847 km above the equator, where the orbital speed is synchronized with the earth's rotation: 23 hours, 56 minutes, 4 seconds.

8.6.2   Navigation Equations

Dork [10] presents the Navstar solution for position location from satellites that radar engineers may find useful. A set i satellites and:

$$(x_i - u_x)^2 + (y_i - u_y)^2 + (z_i - u_z)^2 = (r_i - C_b)^2 \tag{8.8}$$

can be used to solve for a user position. In the above equation x_i, y_i, and z_i are the orthogonal distances of the ith navigation satellite from the center of the earth and $u_{x,y,z}$ are the rectilinear distances of the user's position from the z-y, z-x, and x-y planes respectively. The z-y plane is aligned with the Greenwich meridian; r_i is the radial distance between the ith satellite and the user's position; C_b is a clock bias.

The user's latitude is:

$$\theta = \arccos \frac{\sqrt{u_x^2 + u_y^2}}{|u|} \tag{8.9}$$

The user's longitude is:

$$\phi = \arctan \frac{u_x}{u_y} \tag{8.10}$$

8.6.3   Aircraft Flight

Mach number

Mach number is used to represent the speed of fast airborne vehicles relative to the speed of sound. It is a value not easily derived, however, because it depends upon local temperature, pressure, and air flow characteristics as represented by Reynolds number per unit length of the aircraft. Radar applications do not need exact values of Mach number, so a simple relationship expressing speed of sound as a function of only altitude should suffice:

$$c_s \approx 341 - 4.1h \,, \mathrm{m/s} \quad \text{for } 0 \le h < 11 \text{ km}$$
$$\approx 296 \qquad\qquad \text{for } 11 < h \le 30 \text{ km}$$

where c_s is the speed of sound and h is the aircraft altitude.

Bibliography

[1] Federal Communications Commission, 47 CFR Ch.1 (10-1-86 edition).

[2] IEEE Standard 521-1976, IEEE Standard Letter Designations for Radar-frequency Bands, Reaffirmed 1990.

[3] AFR 55-44, AR 105-86, OPNAVINST 3430.9B, and MCO 3430.1

[4] Johnston, S.L., "Radar frequency management and the new mm-wave radar operating frequencies," Microwave Journal, December 1984, pp. 36-44.

[5] Hovanessian, S.A., "Microwave, millimeter-wave and electro-optical remote sensor systems," Microwave Journal, State-of-the-art Issue, September 1990, pp. 109-129.

[6] Volk, W., *Applied Statistics for Engineers*, McGraw-Hill, New York, 1969, p. 377.

[7] Bishop, L.R., *Approximate standard deviations*, based on Snedcor, G.W., *Statistical Methods*, Iowa State University Press, 1956, p. 44.

[8] Holt, S.M., "RT&E for the next generation weather radar (NEXRAD) system," IEEE AES systems magazine, vol. 4, no. 4, April 1989, pp. 14-18.

[9] Tsandoulas, G.N., "Frequency tradeoffs in space based radar," Military Microwaves, 1986.

[10] Dork, R.A., "Satellite navigation systems for land vehicles," IEEE AES magazine, May 1987, pp. 2-5.

[11] Williams, F.C., personal notes, 1985.

Notes

Notes

Chapter 9

Laser Radar

9.1 Introduction

Because of the sometimes unigue characteristics of topics associated with laser radar, sections of this chapter duplicate those of other chapters. For instance the subjects of detection, clutter, signal processing and propagation are presented here for specific application to laser radar. Chapters on these same subjects in other parts of this book apply to the more general and prevelant radar applications.

9.2 Direct Detection

9.2.1 Signal-to-Noise Ratio

Direct detection is a noncoherent detection process in which the the optical signal is envelope detected in an optical detector that produces an electrical signal. The power signal-to-noise ratio that is background noise limited is, [1]:

$$SNR = \frac{\eta P_S^2}{2h\nu B(P_S + P_B)} \tag{9.1}$$

where η is the detector quantum efficiency (efficiency of converting photons to photo-electrons), P_S is the signal power, P_B is the background noise power, h is Plank's constant ($6.626 \times 10^{34} J \cdot s$), ν is the photon frequency, and B is the detector bandwidth.

From Hovanessian [2], the radar equation 9.2 for a laser radar is:

$$P_S = \frac{8P_T \sigma \rho A^2 \tau_O \eta_L}{\pi^3 R^4 \lambda^2} \tag{9.2}$$

where

P_T is the transmitted power

σ is the laser radar target cross section

ρ is the target backscattering coefficient

A is the aperture area

τ_O is the optical efficiency

η_L is the combined atmospheric and system transmission efficiency

R is the target range

λ is the wavelength

The target cross section, σ, may be smaller than the laser beam, for which, [3]:

$$\sigma = \frac{\pi R^2 \theta_T^2}{4} \tag{9.3}$$

where θ_T is the transmit beamwidth.

The background noise power is:

$$P_B = \sqrt{A_d B}/D^* \tag{9.4}$$

where A_d is the detector area, B is the receiver bandwidth, and D^* is the detector detectivity, [2]. An approximation for the ideal detector detectivity based on data in [3] for a 2 steradian field of view and a 300K background is:

$$D^* \approx A \exp[(\ln \lambda - C)^2/E], cm - Hz^{1/2}/W \tag{9.5}$$

where:

$$A = 2.467 \times 10^{11} \text{ for ideal photoconductor}$$
$$= 3.486 \times 10^{11} \text{ for ideal photovoltaic}$$
$$C = 2.7591$$
$$E = 0.7879$$

Actual detector detectivity values vary differently with frequency and may have values approaching 1/2 to 1/5 of the ideal.

Probability of detection

P_d for direct detection as limited by background noise for a ratio $\sigma_b/\sigma_t = 1$ is shown in Figure 9.1. σ_b is the Gaussian beamwidth and σ_t is the Gaussian uncertainty distribution width of the target location.

Lisko and Paranto, [1] consider the problem of what beamwidth should be used for minimum laser energy if the point target location uncertainity is Gaussian distributed. Their data are shown in Figure 9.2 with the data points. The approximating function for the ratio of rms beamwidth to rms target location uncertainity is:

$$\sigma_b/\sigma_t \approx a \exp[(P_d - b)^2/c], \text{ for } 0.88 < P_d \leq 0.99 \tag{9.6}$$

$$a = 1.4526$$
$$b = 0.8933$$
$$c = 0.0206$$

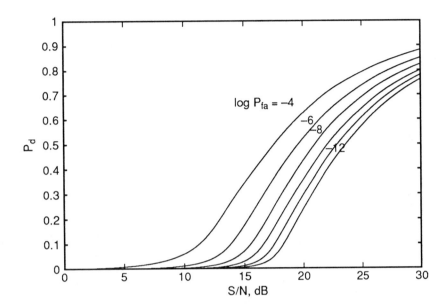

Figure 9.1: The plotted curves are for log $P_{fa} = -4$, -6, -8, -10, -12, and -14 starting at the higher values of P_d as shown.

9.3 Coherent Detection

9.3.1 Signal-to-Noise Ratio

$$SNR = \frac{\eta P_s}{h\nu B} \tag{9.7}$$

where the terms are the same as those defined for the direct detection SNR equation.

The radar equation for a coherent system is the same as in equation 9.2. For coherent detection the noise power is:

$$P_n = \frac{h\nu B}{\eta} \tag{9.8}$$

9.3.2 Probability of Detection

P_d for coherent detection as limited by shot noise for a ratio $\sigma_b/\sigma_t = 1$ is shown in Figure 9.3. σ_b is the Gaussian beamwidth and σ_t is the Gaussian uncertainty distribution width of the target location.

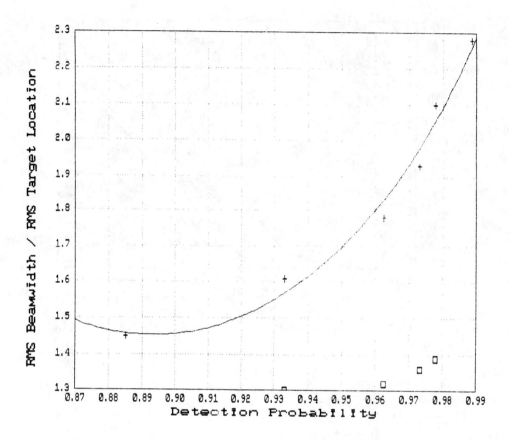

Figure 9.2: Ratio of rms beamwidth to rms target location uncertainity for minimum laser energy for a given probability of detection. The + data apply to direct detection, the □ data apply to coherent detection. The curve follows equation 9.6. Data after [1].

9.4   RF modulation

9.4.1   Power Conversion Efficiency

Kachelmyer and Eng [4] indicate that the ratio of the optical output power to input power for an acousto-optic modulator is:

$$\frac{P_{\text{out}}}{P_{\text{in}}} = \frac{\pi}{2}\sin^2\left(\sqrt{\frac{P_{rf}}{P_{100}}}\right) \tag{9.9}$$

where P_{rf} is the rf drive power, and P_{100} is the drive power where 100% efficiency can be assumed.

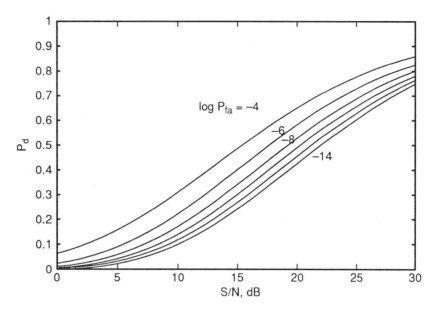

Figure 9.3: P_d for coherent detection as limited by shot noise for $\sigma_b/\sigma_t = 1$, after [1]. The plotted curves are for log $P_{fa} = -4$, -6, -8, -10, -12, and -14 starting at the higher values of P_d as shown.

9.5   Signal Processing

9.5.1   Range Resolution and Doppler Sensitivity Tradeoff

To avoid aliasing of a Doppler shifted target return when using an up and down chirped fm signal the constraint:

$$2f_D + B \leq B_s \qquad (9.10)$$

must be satisfied, where f_D is the Doppler frequency, B is the waveform bandwidth, and B_s is the instantaneous processing bandwidth. Figure 9.4 shows the tradeoff between the waveform range resolution $(c/2B)$ and the maximum allowable target velocity $(\lambda f_D/2)$ for no Doppler ambiguities. Each scale has been normalized with B_s as shown.

9.6   Propagation

Typically atmospheric absorption of laser signals is expressed as a transmittance in %. Brookner [6] and Johnson, [7] present an extensive set of data for various propagation heights, angles and atmospheric consituents. The reader is referred to these data and their references for specific detailed absorption values. Most data are presented for a specific distance, so one must be careful to properly relate such data to the distances of their application.

Although the far IR region (wavelength of $10\mu m$) is a well known window, atmospheric attenuation there is dominated by water vapor absorption. Gramenopoulos [8] illustrates this fact with Figure 9.5.

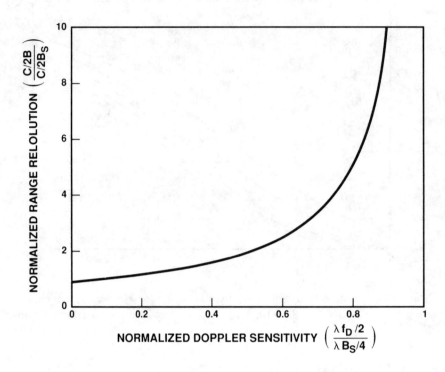

Figure 9.4: Range resolution and Doppler sensitivity tradeoff, from Kachelmyer [5].

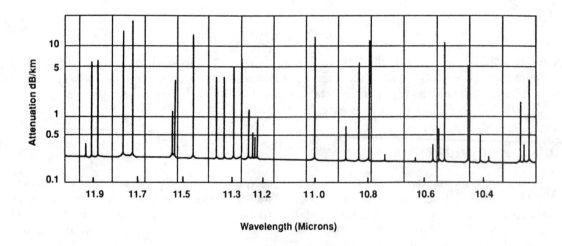

Figure 9.5: Water vapor absorption lines in the far IR region, from Gramenopoulos [8].

Bibliography

[1] Liskow, D.H., J.N. Paranto, "Optimization of laser beam spread to maximize laser radar probability of detection," Proc. SPIE, Vol. 1103, March 29-30, 1989, pp.83- 91.

[2] Hovanessian, S.A., "Microwave, millimeter-wave and electro-optical remote sensor systems," Microwave Journal State of the Art Reference, September 1990, pp. 109-129.

[3] Jelalian, A., "Laser radar theory and technology," *Radar Technology*, E. Brookner, ed., Artech House, Norwood, MA, 1977.

[4] Kachelmyer, A.L., R.S. Eng, "Noise in an acousto-optic modulated laser source," SPIE Proc. Laser Radar 4, R.J. Becherer, ed., March 29-30, 1989, pp. 92-108.

[5] Kachelmyer, A.L., "Laser radar acquisition and tracking," SPIE Proc. Laser Radar 4, R.J. Becherer, ed., March 29-30, 1989, pp. 120-138.

[6] Brookner, E., "Effects of the atmosphere on laser radars," *Radar Technology*, E. Brookner, ed., Artech House, Norwood, MA, 1977.

[7] Johnson, C.M., "Laser Radars," Chap. 37 of *Radar Handbook*, M. Skolnik, ed., McGraw-Hill, 1970.

[8] Gramenopoulos, N., "Concept of a MTI search Ladar," SPIE Proc. Laser Radar 4, R.J. Becherer, ed., March 29-30, 1989, pp. 139-150.

Notes

Notes

Chapter 10

Measurement

10.1 Ambiguities in Measurement

10.1.1 Velocity and Range Ambiguities

The maximum unambiguous velocity measured by a Doppler radar is:

$$v_a = \pm \lambda/4T_s \tag{10.1}$$

and the unambiguous range is:

$$r_a = cT_s/2 \tag{10.2}$$

And also:

$$
\begin{aligned}
r_a &= \frac{c\lambda}{8v_a} \\
\text{or} \\
v_a &= \frac{c\lambda}{8r_a}
\end{aligned}
$$

where T_s is the repetition time of measurement, and c is the propagation velocity. The combination of the two expressions indicates increasing the value of one ambiguous dimension decreases the other, [2]:

$$v_a r_a = \lambda c/8 \tag{10.3}$$

However, with another measurement, in which T_s is changed (staggered), the product can be increased, [2]:

$$v_a r_a = \frac{\lambda c}{8} \frac{1+K}{1-K} , K = T_{s1}/T_{s2} , T_{s2} > T_{s1} \tag{10.4}$$

where T_{s1} and T_{s2} are the two different repetition times.

A necessary condition to maintain an echo sample correlated is:

$$\frac{c\lambda}{4r_a} \geq 2\pi\sigma_v \tag{10.5}$$

239

where σ_v is the Doppler velocity standard deviation.

The total number of range and velocity ambiguities is, [3], p.5:

$$N_{v\&r} = T_c/\tau_c \tag{10.6}$$

where T_c is the coherent processing interval, and τ_c is the compressed pulse width.

10.2   Angle Measurement

10.2.1   Antenna Pattern Measurement

Refer to chapter 1, Measurement criteria, for a discussion of measurement of antenna patterns.

10.2.2   Angle Measurement Accuracy of Conventional Tracking and Search Radars

The prediction of the rms angle accuracy of conventional tracking and search radars is obtained from:

$$\sigma_\theta = K_\theta K_3 \lambda / D \tag{10.7}$$

where K_θ is term that depends upon the implementation used to determine target angle and the SNR of the target return, K_3 is a term that determines the half-power beamwidth of an antenna, and D is the dimension of the antenna in the plane of the angle measurement. From [3]:

$$\begin{aligned} K_3 &\approx 0.72 + 0.0134 g_{sl}, \text{for a rectangular aperture} \\ &\approx 0.86 + 0.0105 g_{sl}, \text{for an elliptical aperture} \end{aligned} \tag{10.8}$$

where g_{sl} is the mainbeam gain relative to the first sidelobe level, dB. The approximation is for $15 \le g_{sl} \le 60$. From [10]:

$$K_\theta = \frac{1}{k_s \sqrt{2(S/N)n_e}} \tag{10.9}$$

where k_s is the error slope at the position of the target position within the antenna beam, (S/N) is the integrated signal-to-noise ratio, and n_e is the number of independent samples. The factor k_s depends upon the angle measurement technique. The reader is referred to [18], chapter 18, and [10], chapter 8, for detailed discussion of k_s. For example a conical scan system $k_s \approx 1.5$; for a scanning antenna system $k_s \approx 2$ and for a monopulse system, [10], figure 8.45:

$$k_s \approx 2.5 \exp(-6.7/g_{sl}) \tag{10.10}$$

We note that, independent of S/N effects, that $K_\theta K_3$ is approximately constant for $13 \le g_{sl} \le 55$.

Effect of target rcs fluctuations on accuracy

Swerling and Peterman, [9], present a general formula for predicting the effect of target fluctuation on rms measurement accuracy as:

$$\sigma = \frac{\rho k}{\sqrt{nx}} \tag{10.11}$$

where n is the number of statistically independent noncoherent signal plus noise samples entering into measurement, x is the per sample S/N, k is a constant depending upon the type of measurement, the waveform, or the antenna beam pattern, and the type of measurement implementation and processing. ρ is a constant depending upon the RCS fluctuation.

For a Swerling 3 target:

$$\rho = \sqrt{2/(1 + 2x_0/\bar{x})} \tag{10.12}$$

For a Swerling 1 target:

$$\rho \approx \sqrt{\ln(\bar{x}/x_0)}, \text{ when } x_0/\bar{x} < 0.001$$
$$\rho \approx 0.8 - 0.25\ln(x_0/\bar{x}), \text{ when } 0.001 \leq x_0/\bar{x} \leq 1$$

where $\bar{x}$ is the average value of x, and x_0 is the signal-to-noise ratio established for target detection.

Barton, [10] pp. 410-412, discusses the additional error caused by target fluctuation where it is found that the time constant of the monopulse normalization circuit has a significant effect on error.

10.2.3 Beam Pointing Accuracy of Wideband Arrays

The angular error obtained with a wideband array used over a wideband frequency relative to an array used over a narrowband frequency is [5]:

$$\sigma_{WB} = \sqrt{L_D}\sigma_{NB} \tag{10.13}$$

where σ_{WB} is the rms angle error for a wideband application, σ_{NB} is the rms angle error for a narrowband application, and $L_D > 1$ is the dispersion loss of a wideband array. The dispersion loss is the loss in mainbeam gain due to the shift in the mainbeam angle from the steered angle, θ_o, caused by an array sensitivity to frequency changes. Frank [4] has computed the energy loss for pulsed signals sent through an array. His data shows the loss of energy is:

$$L \approx 3.4(\Delta\theta_o/\theta_o)^2, \text{ dB}, one-way \tag{10.14}$$

where $\Delta\theta_o/\theta_o < 0.6$ is the beam shift at the spectrum edge expressed in terms of the half-power beamwidth. The shift in beamwidth for a corporate fed array is [4]:

$$\Delta\theta_o \approx -\frac{\Delta f}{f}\frac{1}{\tan\theta_s}, \text{ radians} \tag{10.15}$$

where $\Delta f/f$ is the fractional bandwidth, and θ_s is the beam steering angle from broadside.

10.3 RCS Measurement

10.3.1 Sampling Space for RCS Measurement

The angle increment necessary for RCS measurement is [6]:

$$\Delta\theta \leq \lambda/2D \tag{10.16}$$

and

$$\Delta f \leq \frac{c}{2(1+k)D} \tag{10.17}$$

where $\Delta\theta$ is the sampling angle, Δf is the sampling frequency, D is the largest dimension of the target, and k is a safety factor ≈ 1.

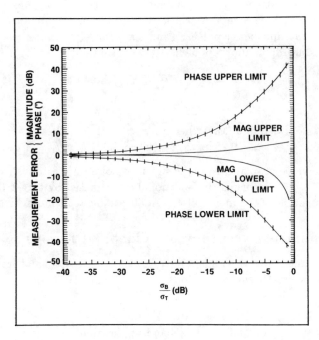

Figure 10.1: Accuracy of RCS measurements as a function of background level relative to target level, from [7].

10.3.2   RCS Measurement Accuracy Dependence on Background Level

The total measured cross section is [7]:

$$\sigma = \sigma_M + \sigma_B + \sigma_T + 2\sqrt{\sigma_B \sigma_T}\cos\phi + 2\sqrt{\sigma_M(\sigma_B + \sigma_T + 2\sqrt{\sigma_B \sigma_T}\cos\phi)}\cos\phi_1 \qquad (10.18)$$

where ϕ_1 is the angle between the resultant target/background vector and the multipath vector σ_M is the multipath cross section of the multipath signal reflected from the target via the chamber wall and from the wall via the target, σ_B is background cross section level, σ_T is the target cross section, ϕ is the angle between the background and target field vectors. The effect of the relative background cross section on RCS measurement accuracy is shown in Figure 10.1. The effect of the relative multipath cross section is shown in Figure 10.2.

10.3.3   RCS Measurement Accuracy Criteria

Welsh and Link, [8] determined the amount of degradation of RCS measurement accuracy due to the amplitude taper of an illuminating beam. For a Gaussian beamshape and a uniformly scattering target, their results for the ratio of measured average RCS to actual average RCS, as corrected, are:

$$\frac{\sigma_{\bar{m}}}{\sigma_{\bar{a}}} = \sqrt{\frac{\pi}{8\ln 2}} R_n \mathrm{erf}(R_n\sqrt{2\ln 2}) \qquad (10.19)$$

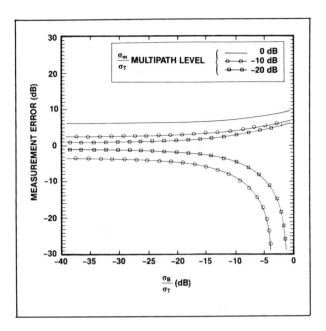

Figure 10.2: Upper and lower error limits for three multipath levels as a function of background level. The zero dB multipath level allows complete cancellation and therefore the lower error limit for this case is negative infinity, from [7].

where R_n is a normalized range to the target.

$$R_n = \frac{R\theta_3}{L} \qquad (10.20)$$

where R is the actual range, θ_3 is the half-power beamwidth, and L is the target projected cross-range length. An appoximation for the measurement degradation is:

$$10\log(\sigma_{\bar{m}}/\sigma_{\bar{a}}) \approx [-0.4849(R_n - 0.1423)^2 - 0.2036]^{-1} , \mathrm{dB} \qquad (10.21)$$

For a measurement error of 1 dB, $R_n = 1.42$.

10.4   Range and Time Measurement

10.4.1   Range Correction

Robertshaw, [11] and [12], using measured and ray trace data from ETAC of the Air Force, explained further in Chapter 15, has created a general expression for the range correction to be applied to measured radar range of an airborne radar:

$$R_c \approx 0.42 + 0.0319 R_t \sqrt{N_s/H_r} , m \qquad (10.22)$$
$$\text{where } 40 \leq R_t \leq 200km$$
$$4.5 \leq H_r \leq 20km$$

The radar altitude, H_r is in kft above sea level, and refractivity, N_s is in N-units. The target altitude is assumed at 1 km above sea level.

Barton, [10] p.306-307, gives an expression for a ground-based radar application:

$$R_c \approx (0.0072 N_s / \sin E)(H_t + 0.33 H_t^2)/(15 + H_t + 0.33 H_t^2) \qquad (10.23)$$

where the elevation angle $E > 3^o$ for R_c accurate to about 5%

10.4.2   Variance of Time Delay Estimation

The minimum variance of time delay estimation, [13], is:

$$\sigma_{TDE}^2 = \frac{3}{8\pi^2} \frac{1 + 2SNR}{SNR^2} \frac{1}{B^3 T} \qquad (10.24)$$

where SNR is the input signal-to-noise ratio of each sample of a coherent signal of bandwidth, B, and T is the integration time of the measuring process. The TB product is large, > 100.

10.4.3   Ranging with Pseudonoise

The maximum range without ambiguity using a pseudonoise ranging system is, [14]:

$$R_p = c\Delta(2^n - 1)/2 \qquad (10.25)$$

where Δ is the chip duration of the pseudo noise sequence, n is the number of feedback shift register stages, and c is the propagation speed.

10.4.4   Multipath Delay Profiles

For an airborne application, Figure 10.3 shows the time difference between a direct path to a target and a single bounce from target-to-ground-to-radar path. The data are presented for a radar altitude of 9.1 km and a 4/3 earth ($a_e = 8495$ km). For a double-bounce multipath, the times shown should be doubled.

The curves shown were computed from equations presented by Fishback, [16], and modified by Blake, [17], starting with the assumption for range to target, R_d, radar altitude, h_1, and target altitude, h_2 and following the sequence:

1. The angle at the center of earth between target and radar is:

$$\sin\phi = 2\arcsin\sqrt{\frac{R_d^2 - (h_2 - h_1)^2}{4(a_e + h_1)(a_e + h_2)}}$$

2. The ground range between the nadir points of h_1 and h_2 is;

$$G = a_e\phi$$

$$p = \frac{2}{\sqrt{3}}\sqrt{a_e(h_1 + h_2) + (G/2)^2}$$

$$\xi = \arcsin\frac{2a_e G(h_2 - h_1)}{p^3}$$

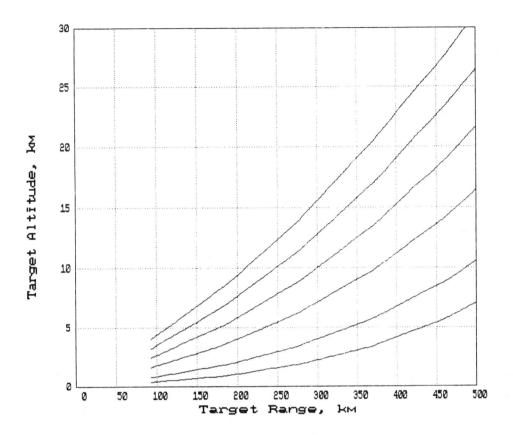

Figure 10.3: Multipath delay profiles shown, starting at the top, 3, 2, 1.5, 1, 0.5, and 0.25μs. Radar altitude of 9.14 km and 4/3 earth ($a_e = 8495$ km) assumed.

3. The range along the ground from bounce point to nadir of h_1 is:

$G_1 = \frac{G}{2} - p\sin(\xi/3)$

4. The range along the ground from bounce point to nadir of h_2 is:

$G_2 = G - G_1$

5. The angle at earth center between the bounce point and target/radar is:

$\phi_i = G_i/a_e$, $i = 1, 2$

6. The range from the target/radar to the bounce point is:

$R_i = \sqrt{h_i^2 + 4a_e(a_e + h_i)\sin^2(\phi_i/2)}$

7. The range difference is:

$\delta = R_1 + R_2 - R_d$

10.5   Noise Figure

10.5.1   Measuring System Noise Figure

Shrader [15] presents a procedure for measuring the effective noise temperature of a phased array radar system using standard data on atmospehric loss, [1]. His procedure is one of using, for example, the following equation to solve for the system noise figure (NF) by measuring the change in system noise for various mainbeam elevation angles. Subsequent configurations at different elevation angles are used to solve for system effective noise temperature to monitor for NF or rf system degradations. A relationship for system effective noise temperature is:

$$T_e = (NF - 1)290 + \frac{3}{L_a} + (1 - \frac{1}{L_a})260 \qquad (10.26)$$

where NF is the sytem noise figure, L_a is the atmospheric loss, 3 represents the cosmic radiation (K), and 260 represents the atmospheric temperature (K). A more extensive equation that includes the effects of antenna sidelobes may also be used.

Bibliography

[1] Skolnik, M., ed., *Radar Handbook*, McGraw-Hill, New York, NY, 1970.

[2] Zrnić, D.S., P. Mahapatra, "Two methods of ambiguity resolution in pulse Doppler weather radars," IEEE Trans. AES-21, no. 4, July 1985, pp. 470-483.

[3] Morchin, W.C., *Airborne Early Warning Radar*, Artech House, Norwood MA, 1990.

[4] Frank, J., "Bandwidth criteria for phased array antennas," *Phased Array Antennas*, A.A. Oliner, G.H. Knittel, eds., Artech House, Norwood, MA, 1972, pp. 243-253.

[5] Jaska, E.A., J.T. Nessmith, L.E. Corey, "Investigations of beam pointing accuracy in wideband phased-array radar," Military Microwaves, July 11-13, 1990, Conf.proc. pp. 545-550.

[6] Fok, F., J. Young, "Space-frequency sampling criteria for electromagnetic scattering of a finite object," IEEE Trans. AP-35, no. 8, Aug. 1987, pp. 920-925.

[7] Matyas, G.J., B.J. Kelsall, "Calibration accuracy considerations for radar cross-section measurements," Microwave Journal, Mar. 1991, pp. 124-132.

[8] Welsh, B.M., J.N. Link, "Accuracy criteria for radar cross section measurements of targets consisting of multiple independent scatterers," IEEE Trans. AP-36, no. 11, November 1988, pp. 1587-1593.

[9] Swerling, P., W.I. Peterman, "Impact of target RCS fluctuations on radar measurement accuracy," IEEE Trans. AES-26, no. 4, July 1990, pp. 685-686.

[10] Barton, D.K., *Modern Radar System Analysis*, Artech House, Norwood, MA, 1988.

[11] Robertshaw, R.G., "Correction to measured range," Microwaves & RF, March 1986.

[12] Robertshaw, R.G., "Range corrections for airborne radar-a Joint STARS study," The Mitre Corp., Bedford MA, project 6460, contract no. F19628-82-C-0001, Air Force Electronic Systems Div., ESD-TR-84-169, May 1984.

[13] Carter, G.C., "Coherence and time delay estimation," Proc. IEEE, February 1987, pp. 236-254.

[14] Yamamoto, Z., et al. "Dual speed PN ranging system for tracking of deep space probes," IEEE Trans. AES-23, no. 4, July 1987.

[15] Shrader, W., "Utilizing atmospheric attenuation to determine system noise temperature of phased array radars," Radar conference, Paris-89, ISSN 0722-8244, vol. 15, no. 2, April 24-28.

247

[16] Kerr, D.E., ed., *Propagation of Short Radio Waves*, M.I.T. Radiation Laboratory Series, v. 13, New York, McGraw-Hill, 1951.

[17] Blake, L.V., *Radar Range-Performance Analysis*, Artech House, Norwood, MA, 1986.

[18] Howard, D.D., "Tracking Radar," Ch.18, *Radar Handbook*, Skolnik, M.I., ed., McGraw-Hill, New York, 1990.

Notes

Notes

Chapter 11

Meteorological Radar

11.1 Introduction

Meteorological radar is a speciality that encompasses many of the topics covered in other chapters. For instance the chapters on clutter, propagation, and Doppler, contain information pertinent to meteorological radar. The possible future sections of this chapter dealing with polarization, radiometry, and profilers are not included.

Some of the prominent sources of detailed papers on the subject can be found in the proceedings of the International Geoscience and Remote Sensing Symposiums, the Conferences on Radar Meteorology, and the IEEE Transactions on Geoscience and Remote Sensing and Radio Science.

A recent book by Bogush [1] is also available in which the author covers the subject in general in good fashion and presents data primarily, but not exclusively, for frequencies above 10 GHz. He also includes a long list of references to many other books and papers. Zrnič and Doviak [2], closely associated with the US NEXRAD radar, have also authored a book on the topic. These same two authored a chapter in the book *Aspects of Modern Radar*, [3], and along with Sirmans authored a paper reprinted in the book *Radar Applications* [4].

11.2 General

11.2.1 Radar Equation for Meteorological Radar

From Serafin, [6], and Bogush [1] the standard radar equation for received power is:

$$P_r = \frac{P_t G^2 \theta \Phi c \tau \pi^3 \mid K \mid^2 Z}{512(2 \ln 2) r^2 \lambda^2 L L_p} \tag{11.1}$$

P_r is the average received power

P_t is the peak transmit power

θ is the azimuth beamwidth

Φ is the elevation beamwidth

c is speed of light

τ is effective pulse width

$\mid K \mid^2$ is $\mid (m^2 - 1)/(m^2 + 2) \mid^2$

m is the complex refractive index of water or ice values of $\mid K \mid^2$ are shown in Table 11.1 for water

$\mid K \mid^2 \approx 0.2$ for ice

Z is the reflectivity factor, see Table 11.2 for values

$Z = AR^B$

R is the rain rate, mm/hr

r is the radar range

L are the system losses

L_p are the propagation losses

Table 11.1: Values of $\mid K \mid^2$ for Water, after Battan [8].

Frequency	Temperature, °C				Average
	−8	0	10	20	
3		0.9340	0.9313	0.9280	0.931
9.3		0.9300	0.9282	0.9275	0.929
24	0.8902	0.9055	0.9152	0.9193	0.908
48	0.7921	0.8312	0.8726	0.8926	0.847
Average =	0.841	0.900	0.912	0.917	0.904

Table 11.2: Values for A, B in $Z = AR^B$ for widespread rain, after Bogush*, [1].

Frequency (GHz)	Rain rate range, mm/hr		
	0-5	5-20	20-100
3	295, 1.45		
5.5	280, 1.45		
6.4	280, 1.45		
9.3	275, 1.55		
16	330, 1.54	330, 1.54	750, 1.30
24	356, 1.50	356, 1.50	820, 1.15
35	350, 1.32	350, 1.32	780, 0.95
48	240, 1.10	240, 1.10	540, 0.75

* Bogush data extracted from Atlas and Kessler [7].

The above equation assumes that the scatterers fill the radar resolution cell and that a Gaussian beam shape is used.

Typically a microburst reflectivity factor, $dBZ \leq 15dB$. And for Denver, Evans et al, [5], found dBZ varied from 0 to 10 dB.

The scatterer reflectivity is related to $\mid K \mid^2$ and Z by:

$$\eta = \frac{\pi^5}{\lambda^4} \mid K \mid^2 Z \times 10^{-18}, \qquad \text{m}^2/\text{m}^3 \tag{11.2}$$

Serafin [6] points out that A and B also change with the type of rain. He suggests the use of:

1. For rain induced or influenced by hills or mountains: $A, B = 31, 1.71$

2. For Thunderstorms: $A, B = 486, 1.37$

3. For snow: $A, B = 2000, 2$, where R is for melted snow.

11.3 Doppler

11.3.1 Doppler and Range Ambiquity

For there to be no ambiguity in the reporting of either target range,r_t, or velocity, v_t, when a single PRF is used the following relationship must be satisfied, [9] [10]:

$$r_t v_t \leq \lambda c/8 \tag{11.3}$$

When more than one PRF is used and there is a difference between the two, such that $PRF_1 < PRF_2$ then:

$$\begin{aligned} r_t v_t &\leq \frac{\lambda c}{8} \frac{1+k}{1-k} \\ k &= PRF_1/PRF_2 \end{aligned} \tag{11.4}$$

As $k \to 1$ the ability to distinguish target velocity from the noise caused by spectral frequency spreading due to equipment or environmental causes is lost. This places a limit on the achievable maximum value of $r_t v_t$. Zrinić and Mahapatra, [9], found the upper limit of $k = 0.4$ for their weather radar application.

11.3.2 Doppler Velocity Distribution of Storms

Figure 11.1 shows the cumulative probability distribution of three typical storms, [11].

11.3.3 Correlation Time of Weather Features

The autocorrelation coefficient, [11], of three weather features is shown in Figure 11.2. The weather features are reflectivity as described by dBZ, the maximum velocity, and the maximum velocity spread in a given resolution cell. The data are segregated into two ranges of reflectivity: (1) 23 to 35 dBZ, referred to as moderate and (2) 58 to 61 dBZ, referred to as high. In each reflectivity range we find that the autocorrelation for each weather feature decreases with increases in time. Furthermore, for a given time, the autocorrelation coefficient for either velocity or velocity spread will drcrease with

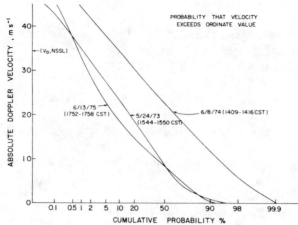

Figure 11.1: Probability distribution of Doppler velocity in three typical storms observed by the US National Severe Storms Laboratory. From Mahapatra and Zrnič [11].

an increasing measure of itself or the other. For instance, the autocorrelation coefficient will decrease with increasing velocity spread for either a fixed velocity or an increasing velocity, but appears to be independent of the reflectivity for these changes. Likewise, the autocorrelation coefficient will decrease with increasing velocity and/or velocity spread but appears to be independent of reflectivity for these changes. For a given time the autocorrelation coefficient will be larger for higher values of reflectivity with an unclear dependence on velocity and velocity spread.

The reader may want to use the data in Figure 11.2 to choose a sampling time for a radar. In general, for aircraft and air traffic information, two samples are required of a given phenomenon. There should be a high correlation between the two samples.

Time to independence for a scanning radar

Eccles [12] and Reid [13] point out that for a scanning radar, assuming a Gaussian beamshape, and a distributed target there are $\sqrt{8\ln 2}$ independent samples in a single coordinate scanning antenna beam. The standard deviation of the velocity estimate is:

$$\sigma_v = \frac{\lambda\sqrt{8\ln 2}}{T4\pi} \tag{11.5}$$

where T is the time it takes for a beam to scan past a point. There are thus $T/\sqrt{8\ln 2}$ independent samples for the above conditions.

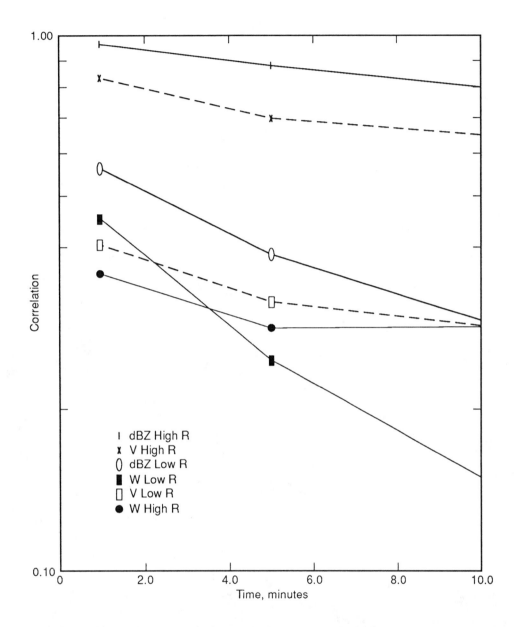

Figure 11.2: Dependence of autocorrelation coefficient on time for three weather features: reflectivity, (dBZ); maximum velocity, shown by the broken line (V); maximum velocity spread, (W). The upper curve for each given weather feature pair pertains to the reflectivity range of 56 to 61 dBZ (high R); the lower to a reflectivity range of 23 to 35 dBZ (low R). Data extracted from Figure 4 of [11].

Bibliography

[1] Bogush, A.J., *Radar and the Atmosphere*, Artech House, Norwood, MA, 1989.

[2] Zrnič, D.S., R.J. Doviak, *Doppler Radar and Weather Observations*, Academic Press, 1984.

[3] Brookner, E., *Aspects of Modern Radar*, Artech House, Norwood, MA, 1988; Chap. 9.

[4] Skolnik, M.I., *Radar Applications*, IEEE Press, New York, 1987; Paper 6.1.

[5] Evans, J.E., W.H. Drury, et al., "Terminal Doppler weather radar clutter model," IEEE Int. Radar Conf., Radar-90, Washington D.C., May 7-10, 1990, pp 12-16.

[6] Serafin, R.J., *Meteorological radar*, Ch. 23 of Radar Handbook, M. Skolnik, ed., McGraw-Hill, New York, 1990.

[7] Atlas, D., E. Kessler III, "A model atmosphere for widespread precipitation," Aeronaut. Eng. Rev., vol. 16, 1957, pp. 69–75.

[8] Battan, L.J., *Radar Observation of the Atmosphere*, Univ. of Chicago Press, Chicago, 1973.

[9] Zrnič, D.S., P. Mahapatra, "Two methods of ambiguity resolution in pulse Doppler weather radars," IEEE Int. Radar Conf., RADAR-85, Arlington, Virginia, May 6-9, 1985, pp. 241–246.

[10] Morchin, W.C., *Airborne Early Warning Radar*, Artech House, Norwood, MA, 1990, p. 237.

[11] Mahapatra, P.R., D.S. Zrnič, "Parameters of a dedicated weather radar for terminal air navigation, traffic control and safety," IEEE International Radar Conference, Arlington, VA, May 6-9, 1985, pp. 235–240.

[12] Eccles, P.J., "The time to independence for a scanning radar," IEEE Int. Radar Conf., RADAR-85, Arlington, VA, May 6-9, 1985, pp. 101–106.

[13] Reid, J.D., "Decorrelation time of weather radar signals," 14th Radar Meteorology Conf., Boston, MA, 1970, pp. 419–420.

257

Notes

Notes

Chapter 12

MTI

12.1 Introduction

There is much material available on MTI systems and performance. The *Radar Handbook*, [1] chapters 16, 17, and some of 18 (1990 ed), The *Space-Based Radar Handbook*, [2], two books by Barton, [3], [4], a book by Nathanson, [5], and a book by Skolnik, [6] are good examples of theory, implementation and performance data on MTI systems. We present some approximating functions here, in some cases published in conference proceedings, that the user of this source book may not be aware of to supplement the above cited references.

12.2 Definitions

A moving target indicator (MTI) is a filter that rejects the clutter sprectrum and passes a broad frequency band that corresponds to target Doppler frequencies. An airborne MTI is denoted by AMTI. Many times though MTI is used in reports and discussions in the place of AMTI. Sometimes AMTI is used for reference to *adaptive MTI*.

Improvement Factor. The MTI improvement factor is defined as:

$$I = \frac{\overline{(S/C)_o}}{(S/C)_i} \tag{12.1}$$

where $(S/C)_o$ is the MTI output of signal to clutter and $(S/C)_i$ is the MTI input signal to clutter. The average is over all target velocities of interest.

Signal-to-Clutter Ratio Improvement or Clutter Attenuation at a Specific Doppler Frequency. In many designs the MTI is followed by a bank of Doppler filters. For this situation the use of *improvement factor* will result in a different value for each Doppler filter. For this application the same ratios are used as for I, except the Doppler frequency corresponding to the target of interest is used rather than averaging over all Doppler frequencies. The $(S/C)_i$ value can be the wideband single pulse value and $(S/C)_o$ can be the narrowband output integration of the filter processed pulses.

Clutter Attenuation. For the MTI system without Doppler filters the clutter attenuation if the ratio of MTI input clutter-to-noise to the output clutter-to-noise.

Subclutter Visibility (SCV). SCV is the maximum $(C/S)_i$ at which the target is still detectable with given probabilities of detection and false alarm in a range gate.

12.3   Finite Impulse Response Filters

12.3.1   Binomial FIR MTI Cancellers

Figure 12.1 shows the improvement factor for the number of pulses, N, processed assuming a Gaussian clutter spectrum. The 3-dB bandwidth is $2.35\sigma_f$, where σ_f is the standard deviation of the Gaussian spectrum. The number of delay stages is $N_d = N - 1$. For the relative clutter spectral spread, $\sigma_f T \leq 0.05$, $(T = 1/PRF)$, the improvement factor can be approximated with:

$$I \approx k/(2\pi\sigma_f T)^{2N_d} \tag{12.2}$$

where $k \approx 3.2369 - 1.7466 \ln N_d$, $2 \leq N_d \leq 6$

$\qquad k = 2$ for $N_d = 1$

For an optimum MTI, in the sense of maximizing improvement factor, the improvement factor follows the same relationship with the k values shown in Table 12.1.

Table 12.1: Values of k for approximating the improvement factor
for an optimum MTI, after Hansen, [7].

N_d	1	2	3	5	8
k	2.131	3.230	3.806	2.516	0.468

The Improvement factor difference between the binomial and optimum coefficient FIR MTI is shown in Figure 12.2, [7].

Cai and Wang, [9], present the approximation for the improvement factor expressed in terms of a user's interference spectrum:

$$I \approx \frac{[1 + (C/N)_i]N}{[1 + (C/N)i\Phi(\Delta f)]} \tag{12.3}$$

The normalized and shifted spectral density of input interference is:

$$\Phi(\Delta f) = \frac{\Phi(f + f_c)}{\Phi(f_c)} \tag{12.4}$$

where $\Delta f = f - f_c$, f_c is the frequency corresponding to the unimodal interference peak and f is the normalized frequency variate. Normalization of f is to the PRF.

$$N >> 1/\mid f - f_c \mid$$

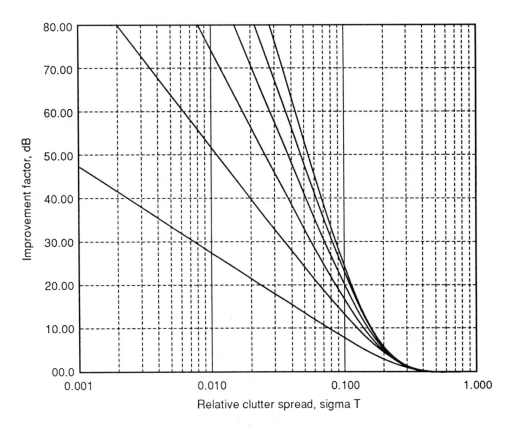

Figure 12.1: Improvement factor of binomial weighted FIR MTI. Relative clutter spectral spread is $\sigma_f T$. The order of the curves starting with the lowest value is N = 2,3,4,5,6, and 7. After Hansen [7] and Skillman [8].

As an example, for the Gaussian clutter spectrum at $f_c = 0$:

$$
\begin{aligned}
\Phi(f) &= \frac{\Phi(f + 0)}{\Phi(0)} \\
&= \frac{\dfrac{P_c}{\sqrt{2\pi}\sigma_c} \exp[-(f + 0)^2/2\sigma_c^2]}{\dfrac{P_c}{\sqrt{2\pi}\sigma_c} \exp(0)} \\
&= \exp(-f^2/2\sigma_c^2)
\end{aligned}
\tag{12.5}
$$

12.3.2 Signal-to-Clutter Improvement for a PD

The signal-to-clutter improvement averaged over all target Doppler frequencies from Hansen, [7], for the optimum pulse Doppler processor is approximated by the above relationship plus N coherent processing intervals. This assumes a Gaussian clutter spectrum $\gg$ noise.

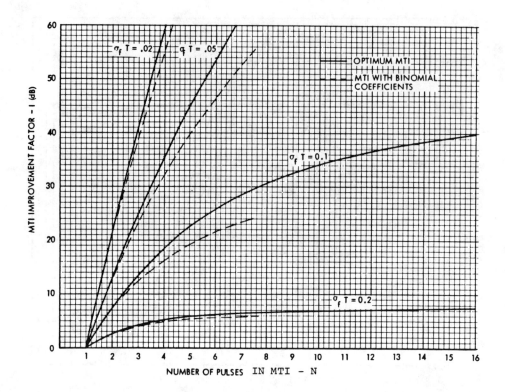

Figure 12.2: Comparison between improvement factor of optimum and binomial coefficient FIR MTIs, from Hansen [7].

The frequency averaged signal gain of the MTI followed by a pulse Doppler processor for a Gaussian clutter spectrum is, [13]:

$$G \approx \frac{2.43 N_c^2}{\pi} [2 \sin(k\pi/N_c)]^{2N_d} \tag{12.6}$$

where N_c is the number of pulses coherently integrated in the Doppler processor, and $k = 0,1,2,\ldots$ $(N_c - 1)$ is a Doppler filter index number. The term $2.43 N_c^2/\pi$ approximates the average power gain for any one of the N_c filters and when $k = N_c/2$ the last half of the above equation containing the sine function reduces to 2^{2N_d} which is the maximum power gain of N_d cascaded delay lines.

12.4   Staggered PRF

12.4.1   A Four-pulse MTI

A four-pulse MTI that uses variable interpulse time periods and variable weights that depend upon the selected time periods is described by Taylor and Brunins, [10]. The patented MTI implemen-

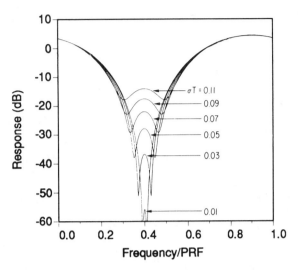

Figure 12.3: Example transfer characteristic for an adaptive 3-pulse MTI canceller.   From Kretschmer, et al. [12].

tation has been produced as an option for the ASRS-3 radar. With T_1, T_2 and T_3 the interpulse periods between and four pulses, and T_1 the oldest, the weights are:

$$\begin{aligned} a &= 1 + 3y/2 + y^2/2 \\ b &= 4 - d - m - k \\ c &= 4 - a - m - k \\ d &= 1 - 3x/2 + x^2/2 \end{aligned}$$
(12.7)

where:

$$\begin{aligned} x &= (T_2 - T_1)/T_2 \\ y &= (T_3 - T_2)/T_2 \\ m &= 4(x - y) + 3xy - x^2y/2 + xy^2/2 - x^2 - y^2 \end{aligned}$$

and k is a small fraction used to move a 0 to a normalized velocity of $(\pi/2)\sqrt{k}$

Shrader and Hansen [11], p.15, provide another algorithm for selecting the time varying weights

12.4.2 An Adaptive 3-pulse MTI

An example 3-pulse adaptive MTI canceller transfer characteric response is shown in Figure 12.3. When normalized by the latent gain, the resulting improvement factor vales in the MTI notch match closely with those values in Figure 12.1 for $N = 3$.

12.5   MTI from a Moving Platform

12.5.1   Compensation for Clutter Doppler Spread

Following the method presented by Hammerle, [13] the improvement factor of an MTI using displaced phase center antenna (DPCA) phase correction for antenna translational motion followed by coherent integration can be approximated by:

$$I_{N_d,N_c,k,x} = I_{N_d,N_c,k,0}/L_{N_D,x,D} \tag{12.8}$$

where N_d is as above, and N_c are the number of coherently integrated pulses following the MTI, $k = 0,1,2,\ldots(N_c-1)$ is a Doppler filter index number, and D pertains to parameter dependence upon DPCA use. $I_{N_d,N_c,k,0}$ is the improvement factor without the use of DPCA and $L_{N_D,x,D}$ is a loss factor resulting from imperfect platform motion compensation by DPCA.

Without DPCA compensation:

$$\begin{aligned} I_{N_d,N_c,k,0} &\approx \frac{2.43}{C_{N_d}\pi}\left[\frac{1}{\pi\sigma_c T}\sin(k\pi/N_c)\right]^{2(N_d+1)} \\ C_{N_d} &= 135\ldots(2N_d+1) \end{aligned} \tag{12.9}$$

The loss factor, $L_{N_d,x,D}$ for 1, 2, and 3 delay MTIs is:

$$\begin{aligned} L_{1,x,D} &\approx 1 + 0.0866(x/\sigma_c T)^2 \\ L_{2,x,D} &\approx 1 + 0.312(x/\sigma_c T)^2 + 0.0123(x/\sigma_c T)^4 \\ L_{3,x,D} &\approx 1 + 0.560(x/\sigma_c T)^2 + 0.079(x/\sigma_c T)^4 + 0.00190(x/\sigma_c T)^6 \\ x &= (V_p T/a)\cos\Phi_e\sin\Phi_a \end{aligned} \tag{12.10}$$

where V_p is the platform velocity
   a is the aperture length
   Φ_e is the antenna beam elevation angle relative to the horizon
   Φ_a is the antenna beam horizontal azimuth angle

12.5.2   DPCA Enhancement of the Improvement Factor

The degradation of I caused by platform motion without use of DPCA, [13], is:

$$L_{N_d,x,\bar{D}} \approx [1 + 0.36(x/\sigma_c T)^2]^{N_d+1} \tag{12.11}$$

The enhancement of I when DPCA is used is $L_{N_d,x,\bar{D}}/L_{N_d,x,D}$. Results for $N_d = 1,2,$ and 3 followed by coherent integration are shown in Figure 12.4.

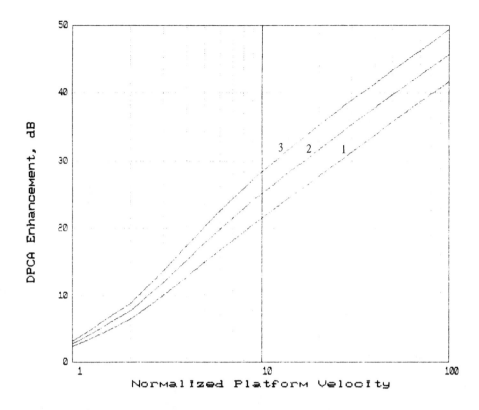

Figure 12.4: DPCA enhancement for 1,2, and 3 delay line MTI followed by coherent integration. Normalized platform velocity is $V = x/\sigma_c T$. (1) is for $N_d = 1$ with $N_c > 4$ or $N_d = 2$ for $N = 1$, (2) is for $N_d = 2$ with $N_c > 4$ or $N_d = 3$ for $N = 1$, and (3) is for $N_d = 3$ with $N_c > 4$ or $N_d = 4$ for $N_c = 1$. After procedures given by Hammerle [13].

Bibliography

[1] Skolnik, M.I., ed., *Radar Handbook*, McGraw-Hill, New York, 1990.

[2] Andrews, G.A., K. Gerlach, "SBR clutter and interference," Ch. 11, *Space-Based Radar Handbook*, Cantafio, L.J., ed., Artech House, Norwood, MA, 1989.

[3] Barton, D.K., *Radar System Analysis*, Artech House, Norwood, MA, 1977, ch.7.

[4] Barton, D.K., *Modern Radar System Analysis*, Artech House, Norwood, MA, 1988, ch. 5.

[5] Nathanson, F.E., *Radar Design Principles*, McGraw-Hill, New York, 1969, ch.9.

[6] Skolnik, M.I., *Introduction to Radar Systems*, McGraw-Hill, New York, 1980, ch.4.

[7] Hansen, V.G., "Topics in radar signal processing," Microwave Journal, March 1984.

[8] Skillman, W.A., *Radar Calculations Using Personal Computers*, Artech House, Norwood, MA, 1984.

[9] Cai, L., H. Wang, "An approximate improvement factor expression in terms of interference spectrum," IEEE Trans. AES-25, no.6, November 1989, pp. 898- 902, and correction Trans. AES-26, no.5, September 1990, p. 899.

[10] Taylor, J.W., G. Brunins, "Long-range surveillance radars for automatic control systems," IEEE Int. Radar Conf., RADAR-75, Arlington, VA, April 21–23, 1976, pp. 312–317.

[11] Shrader, W.W., V. Gregers-Hansen, "MTI Radar," ch. 15 of *Radar Handbook*, M.I. Skolnik, ed., McGraw-Hill, New York, 1990.

[12] Kretschmer, F.F., B.L. Lewis, C. Lin, "Adaptive MTI and Doppler filter bank clutter processing," IEEE Nat. Radar Conference, Atlanta, GA, March 13–14, 1984, pp. 69–73.

[13] Hammerle, K.J., "Cascaded MTI and coherent integration techniques with motion compensation," IEEE Int. Radar Conf., Arlington, VA, May 7–9, 1990, pp. 164–169.

Notes

Notes

Chapter 13

Multistatic/Bistatic Radar

13.1 Introduction

A bistatic radar is a radar system in which the transmitting and receiving antennas are separated by a distance which is large compared with the antenna size. A multistatic radar is a multiplicity of bistatic radars; that is, a system in which there are more than one receiver and/or more than one transmitter. The material we have compiled are directed to the bistatic system.

13.2 General Radar Equation

Willis [6] presents a general equation which can be applied to all types of waveforms. His equation is:

$$(R_T R_R)^2 = \frac{P_T G_T G_R \lambda^2 \sigma_b F_T^2 F_R^2}{(4\pi)^3 K T_s B_n (S/N) L_T L_R} \tag{13.1}$$

where

P_T is the peak transmitter power

G_T and G_R are the transmit and receive antenna gains,

σ_b is the bistatic radar cross section,

F_T is the pattern propagation factor for the transmitter to target path,

F_R is the pattern propagatioin factor for the receiver to target path,

T_s is the receiving system noise temperature,

B_n is the noise bandwidth of the receiver predetectioin filter, sufficient to pass all spectral components of the transmitted signal,

L_T and L_R are the transmitting and receiving system losses not included in the other factors.

273

13.3   Bistatic Search Equation

The radar search equation can be applied to bistatic systems by changing the interpretation of some of the terms, [1], The angular surveillance volume of the monostatic radar equation changes to the bistatic angular volume common to the transmitter and receiver and the effective aperture changes to the aperture of the smaller of either the transmitter or the receiver, depending on which one is used to floodlight the area to be searched. The bistatic equation for search becomes:

$$(R_T R_R)^2 = \frac{P_{avg} A_{fl} t_s \sigma_b}{4\pi k T N F L_b \Omega_b (S/N)_i} \tag{13.2}$$

where

A_{fl} is the effective aperture of a floodlight antenna, which may be the transmit or the receive aperture depending upon which is used to floodlight an area to be searched,

t_s is the time to search the angular search volume,

Ω_b is the angular search volume,

NF is the receiving system noise figure,

L_b are the bistatic radar system losses, and

$(S/N)_i$ is the integrated signal-to-noise ratio.

13.4   Constant Sensitivity Contours

The locus of constant sensitivity are described by the ovals of Cassini [3]:

$$(\rho^2 + 1)^2 - 4\rho^2 \cos^2 \theta - C^4 = 0 \tag{13.3}$$

where $\rho^2 = rA$

r and θ are the values of a circular coordinate system centered at the mid-point between the transmitter and receiver

A is one-half the distance between the transmitter and receiver

$C^4 = (R_T R_R)^2 / A^2$

Use the quadratic equation with $\rho^2 = x$, $a = 1$, $b = 2 - 4\cos^2 \theta$, $c = 1 - C^4$ to solve for r and θ. With the bistatic receiver located at the monostatic range, the bistatic range will be extended 1.62 times the monostatic range along the bistatic axis relative to the transmitter. This statement assumes that the losses, integration time, antenna gains and radar cross section are equal between the two systems. Figure 13.1 shows the change of cross-range and axis-range values with the receiver to transmitter distance as a baseline normalized by the monostatic range. The cross range value shown extends from the baseline outward to one side. It should be doubled to find the surveillance volume. The maximum cross-range value decreases with increasing baseline distance. The maximum cross-range occurs at the mid point of the baseline for transmitter to receiver separations up to $\sqrt{2}R_{mono}$ where R_{mono} is the monostatic detection range. Beyond $\sqrt{2}R_{mono}$ a minima of cross-range occurs

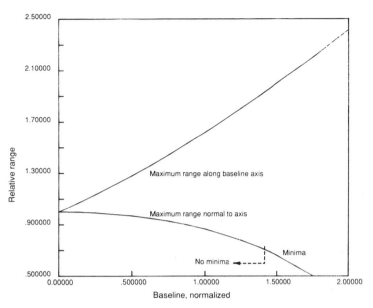

Figure 13.1: Bistatic baseline cross-range and baseline-range relative to a monostatic radar of equal sensitivity, from [4].

at this same baseline mid-point with two maximum points at each side of the baseline mid-point that move toward the transmitter and receiver positions with an increasing baseline. The minima distance is shown plotted in Figure 13.1 for baselines $> \sqrt{2}R_{mono}$. The minima cross-range becomes zero when the baseline is $2R_{mono}$. For baselines $> 2R_{mono}$ the equal sensitivity contour lines encircle the transmitter and receiver in pairs.

13.5   Clutter Cell Area

13.5.1   Beamwidth Limited Clutter Cell Area [6]

$$(A_c)_b \approx \frac{(R_R\theta_R)(R_T\theta_T)}{\sin\beta} \tag{13.4}$$

where

θ_R is the receive half-power beamwidth

θ_T is the transmit half-power beamwidth

β is the bistatic angle which is the angle between the transmitter and receiver with the target as the apex.

13.5.2   Range Limited Clutter Cell Area [6]

$$(A_c)_r \approx \frac{c\tau R_R\theta_B}{2\cos^2(\beta/2)} \tag{13.5}$$

where

θ_B is the beamwidth of the antenna closest to the clutter area of interest and the eccentricity $e < 0.95$.

$$e = A/(R_T + R_R) \tag{13.6}$$

13.6   Unambiguous Range and PRF

The unambiguous range is [6]:

$$(R_T + R_R)_u = c/f_r \tag{13.7}$$

where f_r is the PRF.

$$f_{ru} \approx \frac{c\tan(\beta/2)}{(R_T\theta_T + R_R\theta_R)} \tag{13.8}$$

where f_{ru} is the unambiguous PRF. The approximation breaks down as β approaches $0°$ or $180°$.

13.7   Target Doppler

The doppler frequency caused by target motion, in the case of a stationary transmiter and receiver is [6]:

$$f_{DB} = (2V/\lambda)\cos\delta\cos(\beta/2) \tag{13.9}$$

where

V is the target velocity

δ is the angle between the target velocity vector and the bisector of the bistatic angle

The doppler frequency at the receiver for a stationary target but with both a moving transmitter and receiver is:

$$f_{DB} = (V_T/\lambda)\cos(\delta_T - \phi_T) + (V_R/\lambda)\cos(\delta_R - \phi_R) \tag{13.10}$$

where

V_T is the transmitter velocity

V_R is the receiver velocity

δ_T is the angle between the transmitter velocity vector and the normal to the bistatic baseline where the normal is directed towards the target.

δ_R is similar to δ_T, except the receiver velocity is used

ϕ_T is the angle between the transmitter line-of-sight to the target and the normal to the baseline, where the normal is directed towards the target.

ϕ_R is similar to ϕ_R, except the receiver line-of-sight is used.

13.8 Range Resolution

The physical separation between two targets to be resolved must be [6]:

$$\Delta R_\phi \approx c\tau[2\cos(\beta/2)\cos\phi] \tag{13.11}$$

where

τ is the radar compressed pulsewidth

ϕ is the angle between the bisector of the bistatic angle and a line joining the position of one target located on the bisector to the other target.

13.9 Doppler Resolution

The required velocity difference between two colocated targets to be resolved in Doppler must be [6]:

$$\Delta V = \lambda/[2T_{cpi}\cos(\beta/2)] \tag{13.12}$$

where T_{cpi} is the radar coherent processing time

13.10 Angle Resolution

Two targets must be separated by:

$$S \approx 2\theta_{T/R}R_{T/R}/\cos(\beta/2) \tag{13.13}$$

to achieve a cross-range separation between two targets on the same isorange that is equal to the null-to-null one-way beamwidth of one of the beams [6].

The subscript means that the lower product of $\theta_{(.)}$, $R_{(.)}$ is used for either the transmit or receive beamwidth and range values–that is, if one product is more than twice the other. If the products are equal, the value 2 should be deleted. A transition region is defined between the use of 2 or not in the above equation if one product is within a factor of 2 of the other. A Gaussian curve may be used to transition between the factor of 1 and 2 in the above equation according to the ratio of one product to the other. The above equation also applies to non-isorange situations when $2\theta_{R/T}R_{R/T} >> c\tau[2\cos(\beta/2)]$, the typical case.

13.11 Target Range Determination

We find the range to the target by measuring α_R and ΔR, where ΔR is the range difference between the bistatic path length and the direct path length between transmitter and receiver. α_T is calculated from measurement of the transmitter antenna scan rate and the time of transmitter mainbeam reception at the receiver. α_R and α_T are the angles between the bistatic baseline and the lines-of-sight to the target from the transmitter and receiver, respectively. From [5] for a uniformly scanning transmitter:

$$R_R = \frac{\Delta R \sin\alpha_T}{\sin\alpha_R + \sin\alpha_T - \sin(\alpha_T + \alpha_R)} \tag{13.14}$$

$$R_O = \frac{\Delta R \sin(\alpha_R + \alpha_R)}{\sin \alpha_R + \sin \alpha_T - \sin(\alpha_R + \alpha_R)} \qquad (13.15)$$

where R_O is the baseline distance.

$$R_T = \Delta R - R_R + R_O$$

Bibliography

[1] Ewing, H.F., "The Applicability of Bistatic Radar to Short Range Surveillance," IEE Int. Radar Conf., RADAR 77, London, Oct., 1977, pp. 53-58.

[2] Blake, L.V., *Radar Range-Performance Analysis*, Artech House, Norwood, MA, 1986.

[3] Caspers, J.W., "Bistatic and Multistatic Radar," Ch. 36 in *Radar Handbook*, Skolnik, M.I. (ed.), McGraw-Hill, New York 1970.

[4] Morchin, W.C., *Airborne Early Warning Radar*, Artech House, Norwood MA, 1990, p. 357.

[5] Stein, I., "Bistatic radar applictions in passive systems," Jour. Electronic Defense, Mar. 1990, pp. 55-61.

[6] Willis, N.J., *Bistatic Radar*, Artech House, 1990, Norwood, MA.

Notes

Notes

Chapter 14

Over-the-Horizon Radar

14.1 Introduction

HF over-the-horizon radar (OTH) is a field of radar that does not have as much engineering activity as the other more common radar designs. As a consequence there is not as much technical material available. We present some supplemental material to suggested publications.

The first open tutorial overview of OTH was given by Headrick and Skolnik, [1]. This was followed by papers that provided an understanding of sea clutter, [2] and [9]. The only book devoted to the subject was authored by Kolosov et al. [5]. In addition one can find abundant practical design data in a chapter written by Headrick [4]. One can find the most recent information in technical papers of the IEEE Journal of Oceanic Engineering and the International Geoscience and Remote Sensing Symposia. Specific references are given by Ausherman, [6].

14.2 Target Detection Models

Fante and Dahr, [7], have determined the required signal-to-noise ratio for various types of targets and various combinations of Faraday rotation, multipath, and ionospheric fluctuations. Their results are shown in Figures 14.1 - 14.6. Figure 14.1 shows the required signal-to-noise ratio $(S/N)_0$ for a nonfluctuating target in the absence of Faraday rotation for various types of targets and multipath conditions. Figure 14.2 is similar except there is no multipath but there is Faraday rotation. A symmetric target is one for which the vertically polarized radar cross section is equal to the horizontally polarized radar cross section. Figure 14.3 is for conditions in which there is both Faraday rotation and multipath for a nonfluctuating target. For these data $\rho = \rho_v = -\rho_H$. Figure 14.4 is the same except for a Rayleigh fluctuating target. Figure 14.5 is again the same except for a linearly polarized target. For this figure if the target is horizontally polarized, set $\rho = \rho_H$, if the target is vertically polarized set $\rho = \rho_v$.

Figure 14.6 shows the product of the required signal-to-noise ratio, $(S/N)_0$, and parameters that describe the average ionospheric conditions. These data include the effects of multipath, Faraday rotation and ionospheric fluctuations. Fante and Dhar report that the term in the figure γ_2 which accounts for the average noise level after ionospheric reflection is about 0.6. And, the term γ_1 which accounts for the average ionospheric reflection coefficient is about 0.8.

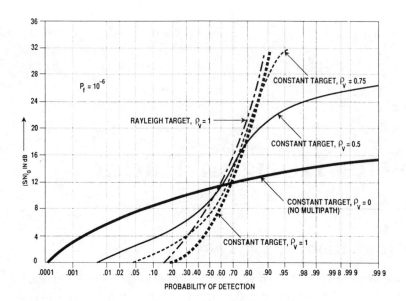

Figure 14.1: Required $(S/N)_0$ for detecting a target in Rayleigh noise with multipath present but Faraday rotation absent, $P_{fa} = 10^{-6}$. ρ_V is the Earth reflection coefficient for vertical polarization. From Fante and Dahr [7].

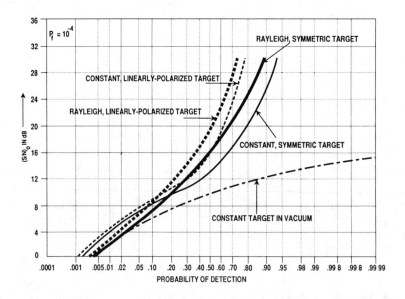

Figure 14.2: Required $(S/N)_0$ for detecting a target in Rayleigh noise with Faraday rotation present but multipath absent, $P_{fa} = 10^{-6}$. From Fante and Dahr [7].

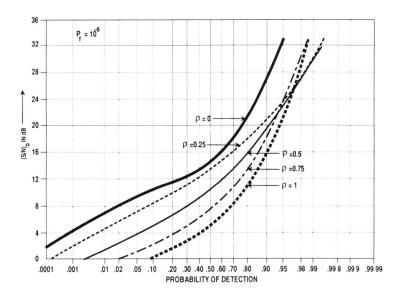

Figure 14.3: Required $(S/N)_0$ for detecting a nonfluctuating symmetric target in Rayleigh noise with multipath and Faraday rotation present, $P_{fa} = 10^{-6}$. ρ is Earth reflection coefficient. From Fante and Dahr [7].

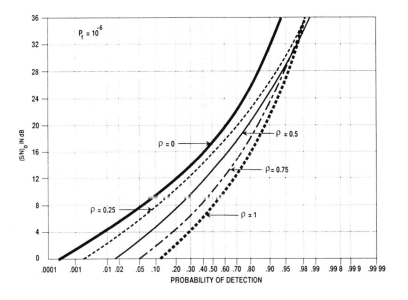

Figure 14.4: Required $(S/N)_0$ for detecting a fluctuating symmetric target in Rayleigh noise with multipath and Faraday rotation present, $P_{fa} = 10^{-6}$. ρ is Earth reflection coefficient. From Fante and Dahr [7].

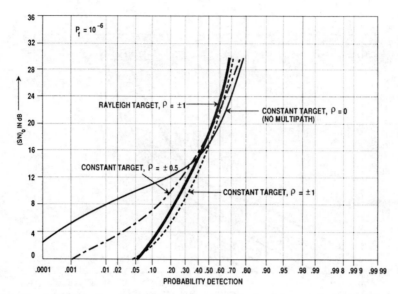

Figure 14.5: Required $(S/N)_0$ for detecting a linearly polarized target in Rayleigh noise with multipath present but Faraday rotation absent, $P_{fa} = 10^{-6}$. ρ is Earth reflection coefficient. From Fante and Dahr [7].

The terms are defined by:

$$\gamma_1^2 = <\ln(K/K_0)>$$
$$\gamma_2^2 = <\ln(\sigma/\sigma_0)> \tag{14.1}$$

where K and K_0 are obtained from:

$$K^2 = (S/N)_0 \; \sigma^2/\sigma_T$$
$$\ln K_0 = <\ln K>$$

and σ is the rms clutter plus noise amplitude, $\ln \sigma_0 = <\ln \sigma>$, and $<\cdot>$ indicates an ensemble average value. The term C in the figure is:

$$C = (\gamma_1^2 + \gamma_2^2 - 2r\gamma_1\gamma_2)^{1/2} \tag{14.2}$$

where r is the correlation coefficient between the signal and noise plus clutter. Fante and Dhar find $C \approx 0.65$.

Required signal-to-noise ratio which includes Faraday rotation and multipath for a symmetric target is:

$$\left(\frac{S}{N}\right) = 0.5(1 + 8\rho^2 + \rho^8)(S/N)_0 \tag{14.3}$$

where $\rho = \rho_v = -\rho_H$ is the Earth reflection coefficient.

14.3   Clutter

14.3.1   Sea Bragg-line Doppler

The Doppler frequency of the first order Bragg sea clutter backscatter is, [8]

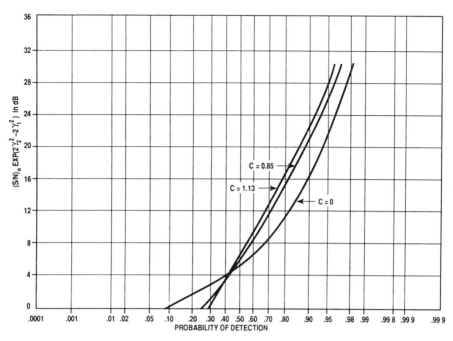

Figure 14.6: Average probability of detection for detecting a symmetric target in Rayleigh noise including the effects of multipath, Faraday rotation and ionospheric fluctuations, $P_{fa} = 10^{-6}$ and Earth reflection coeficient, $\rho = 1$. See text for definition of C. From Fante and Dahr [7].

$$f_{ds} = \pm\sqrt{\frac{gf_o}{\pi c}} \tag{14.4}$$

where g is the acceleration due to gravity, f_o is the radar frequency, and c is the speed of propagation. Radar platform motion will vectorially combine with the Bragg-line Doppler.

14.3.2   Sea/Wind Direction Measurement

Long and Trizna as referred to by Ahearn et al, [9] showed a relationship between the wind direction and the ratio of the first Bragg line sea clutter amplitude returns. They empirically determined that the angle between the wind direction and the radar beam was:

$$\Theta = 0.5\cos^{-1}(-1.1348 + 0.1348\rho') \tag{14.5}$$

where ρ is the ratio of the advancing to receeding first Bragg-line amplitudes. The relationship is shown plotted as the curve in Figure 14.7. The radar beamwidth varied with frequency from $8°$ to $10°$. Shearman et al, [10] later showed the effects of the radar beamwidth upon the angle determination. Their data are shown in Figure 14.7 as the data points.

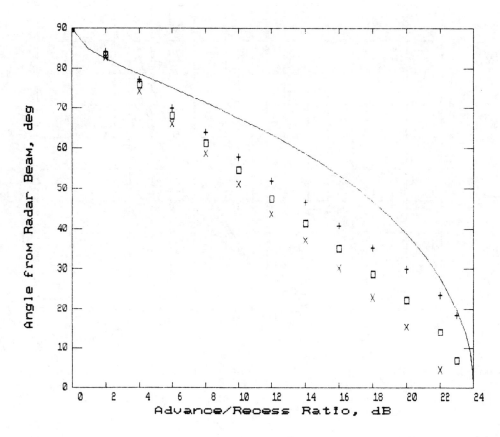

Figure 14.7: Sea/wind direction as determined from the amplitude ratio of the advancing and receeding first Bragg lines. The curve is after Ahearn et al. [9]. The data points are after Shearman et al. [10]. + is for a 10° radar beam, □ is for a 50° radar beam and × is for a 70° radar beam.

Bibliography

[1] Headrick, J.M., M.I. Skolnik, "Over-the-Horizon Radar in the HF Band," IEEE Proc. vol. 62, no. 6, June 1974, pp. 664-673.

[2] Barrick, D.E., et al. "Sea Backscatter at HF: Interpretation and Utilization of the Echo," IEEE Proc., vol. 62, no. 6, June 1974, pp. 673-680.

[3] Ahearn, J.L. et al. "Tests of Remote Skywave Measurement of Ocean Surface Conditions," IEEE Proc. vol. 62, no. 6, June 1974, pp. 681-687.

[4] Headrick, J.M., "HF over-the-horizon radar," Ch. 24 of *Radar Handbook*, M.I.Skolnik, ed., McGraw-Hill, New York, 1990.

[5] Kolosov, A.A., et al., translation by W.F. Barton, *Over-the-Horizon Radar*, Artech House, Norwood, MA, 1987.

[6] Ausherman, D.A., "Cumulative index on radar systems for 1985–1989," IEEE Trans. AES-27, no. 3, May 1991.

[7] Fante, R.L., S. Dhar, "A model for target detection with over-the-horizon radar," IEEE Trans. AES-26, no. 1, January 1990, pp. 68–83.

[8] Ponsford, A.M., D.G. Money, M.H. Gledhill, "Progress in ship tracking by HF ground-wave radar," IEEE Int. Radar Conference, London, 1987, pp. 89–96.

[9] Ahearn, J.L., S. Curley, J. Headrick, D. Trizna, "Tests of remote skywave measurement of ocean surface conditions," IEEE Proc. vol. 62, no. 6, June 1974, pp. 681–687.

[10] Shearman, E.D.R., et al. "HF ground-wave radar for sea-state and swell measurement; theoretical studies, experiments and proposals," Int. Radar Conf. RADAR-82, London, 1982, pp. 101–106.

Additional Reading

General

[11] Headrick, J.M., "Looking over the horizon," IEEE Spectrum, July 1990, pp. 36-39.

[12] Boyle, D. "Radar with its feet in the sea - OTH radar clinging to the surface," International Defense Review, April 1989, pp. 501–503. Includes a description of a Marconi experimental monostatic system using a ground wave mode for surveillance. The article provides relative cost comparisons to airborne early warning radar systems.

Track

[13] Ralph, A.P., "Data processing for a groundwave HF radar," GEC Jour. of Research, vol. 6, no. 2, 1988, pp. 96–105. Discusses various methods of detecting and tracking a maneuvering target.

Anti-interference

[14] Madden, J.M., "The adaptive suppression of interference in HF ground wave radar," Int. Radar Conference, London, 1987, pp. 98–102. Describes a method of adaptive antenna null steering in polarization space.

Notes

Notes

Chapter 15

Propagation

15.1 General

15.1.1 Upper Frequency Limit of Coaxial Cable

The frequency of operation for a coaxial cable in the TEM mode is constrained by [1]:

$$f < \frac{2v}{1.873\pi(a+b)} \tag{15.1}$$

where v is the speed of propagation along a coaxial cable($v = c/\sqrt{\epsilon}$, ϵ is the dielectric constant of material between the inner and outer conductor), a is the inner conductor radius, and b is the outer conductor radius.

15.2 Absorption

15.2.1 Clear Atmosphere

Figure 15.1 shows atmospheric attenuation for the clear atmosphere. The values shown are larger than earlier work, [3] and [4], because of additions to account for the effect of water vapor absorption due to resonance lines above 100 GHz. These resonances above 100 GHz increase the apsorption for tho lower microwave frequencies. The larger values also account for improved theroretical expressions for oxygen absorption and an accounting for the separate contributions of each individual oxygen resonance near 60 GHz, [5].

The reader is referred to Blake, [2] for the integrated path loss for ground-to-air radar applications and to Morchin, [6] for air-to-air radar applications.

Altshuler and Marr [19] in a later work give an approximation based on measurements for the one-way absorption coefficient at frequencies of 15 and 35 GHz between elevation angles of 0 to 20° relative to the horizon as:

$$\begin{aligned} \alpha_{15} &= 0.0045 + 0.0182\rho, \text{ dB/km for 15 GHz,} \\ \alpha_{35} &= 0.0027 + 0.0068\rho, \text{ dB/km for 35 GHz} \end{aligned} \tag{15.2}$$

and ρ is the surface absolute humidity, g/m^3.

293

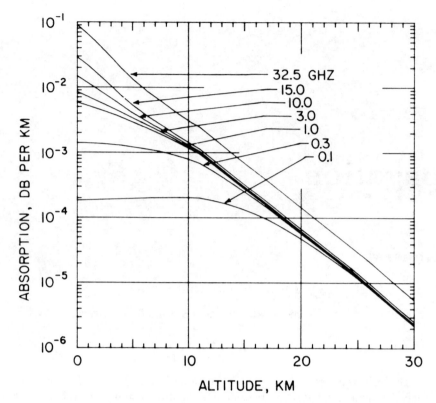

Figure 15.1: Total absorption coefficient (oxygen plus water vapor) in the standard atmosphere. From [2].

Galm et al. [8] reported measurement results that one can approximate with a function for use at 337 GHz:

$$\alpha \approx 3.59 - 0.11T , \ \text{dB/km/[g/m}^3] \tag{15.3}$$

where T° C, is the middle temperature value of a 5° C range. Data were obtained for $15.5 \leq T < 21.5$. $[\,\cdot\,]$ indicates absolute humidity that spanned the range 14 to 19.5 g/m^3 in the tests.

A more frequency extensive set of clear atmosphere specific attenuation data, [9] are shown in Figure 15.2.

Absolute humidity   Absolute humidity, gm/m^3, or water vapor density as given by Tank [39] can be use to approximate the U.S. Standard atmosphere by:

$$
\begin{aligned}
\rho_w &\approx 12.12 \exp(-0.723h) , h \leq 10\text{km} \\
&\approx 11.83 \exp(-0.72h) , 10 < h \leq 13.5\text{km} \\
&\approx 0.0047 \exp(-0.14h), \ , 13.5 < h \leq 30\text{km}
\end{aligned}
\tag{15.4}
$$

where h is height above sea level, km. It is to be noted that the constant 12.12 can be in the range of about 0.12 to 24 depending upon location on earth [39].

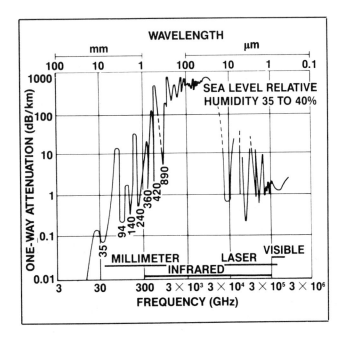

Figure 15.2: Clear atmosphere attenuation versus frequency. From [9].

15.2.2   Rain

A common approximation for rain attenuation, expressed in dB/km for one-way propagation, is of the form [11]:

$$\alpha \approx a(f)R^{b(f)} \tag{15.5}$$

where R is the rain rate, expressed in mm/hr. The approximation is more easily limited to the microwave frequency interval, about 1 to 25 GHz rather than presenting an extensive table of values for $a(f)$ and $b(f)$. The terms

$$a(f) = 4 \times 10^{-5} F_{GHz}^{2.44}$$
$$b(f) = 0.79 \left(0.99^{F_{GHz}}\right) F_{GHz}^{0.16} \tag{15.6}$$

are suggested. The $a(f)$ term does show some dependency on R, particularily in the 100 mm/hr range; $a(f)$ is about 30% low when compared to the $R = 101.6mm/hr$ curve in Figure 15.3.

However, data presented by Manabe et al. [12] for frequecies of 50, 80, 140 and 240 GHz indicate:

$$a(f) = 4 - 4\exp(-0.0037 F_{GHz})$$
$$b(f) = 1.04 - 0.18\log F_{GHz} \tag{15.7}$$

For frequencies above 25 GHz, other approximating equations may be applicable. We find the following to be useful:
For the frequency interval; $25 \le F_{GHz} < 180$:

$$\alpha \approx A(R)\exp[(\ln F_{GHz} - B(R))^2/C(R)] \tag{15.8}$$

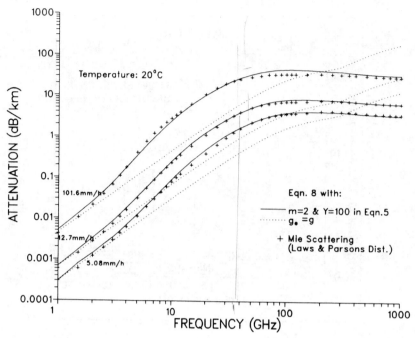

Figure 15.3: Specific rain attenuation in rain. From [18]; see [18] for figure notes.

$$A(R) = 1.15R^{0.73}$$
where $B(R) = 4.55 + 1.59/R$
$$C(R) = -1.44 - 0.0087R$$
In the frequency region 180 to 1000 GHz:

$$\alpha \approx D(R) - E(R)F_{GHz} \tag{15.9}$$

where $\begin{aligned}D(R) &= 1.43R^{0.69} \\ E(R) &= 5 \times 10^{-4}R^{0.534}\end{aligned}$

Attenuation dependence on rain drop size

Sekine, Chen, and Musha, [13] relate the attenuation rate to drop size distribution that in turn depends upon rain rate. Their results are shown in Figure 15.4 for a rain rate of 50 mm/hr for two types of drop size distribution functions; the log-normal and the Weibull. The attenuation rate for the log-normal distributed rain drop size is shown to also depend upon the category of the rain-showers and two levels of thunderstorms as discussed in the caption for Figure 15.4.

The authors [13] propose the use of the Weibull distribution for the density of rain drops as a function of rain drop size for a given rain rate as:

$$N(D) = N_o \left(\frac{c}{b}\right) \left(\frac{D}{b}\right)^{c-1} \exp -(D/b)^c$$

where

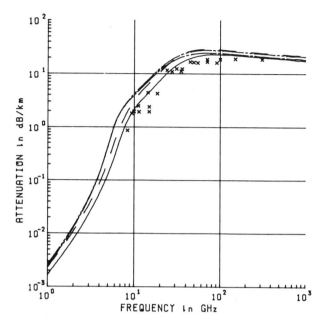

Figure 15.4: Rain attenuation for Weibull and Log- normal distributed rain drop size at 50 mm/hr. Weibull drop size distribution; ——————. Log-normal drop size distribution for showers (5 < R ≤ 50 mm/hr); — — —. Log-normal drop size distribution for thunderstorms (5 < R ≤ 50 mm/hr); — · — · — · . Log-normal drop size distribution for thunderstorms (50 < R ≤ 200 mm/hr); — · · — · · — . From [13].

$$D = 1.238 R^{0.182}, \text{mm}$$
$$N_o = 1000 m^{-3}$$
$$b = 0.26 R^{0.44} \text{mm}$$
$$c = 0.95 R^{0.14} \text{mm}$$

Similarily Manabe et al. [12] use an exponential distribution based on their work on data in the 11.5 to 81.8 GHz frequency range:

$$N(D) = 17300 R^{-0.16} \exp(-5.11 R^{-0.253} D) \tag{15.10}$$

Path length through rain

The length of path through rain depends upon the geometry between the radar and target and the rain distribution and statistics. Manabe et al. [12] provide an empirical relationship for the spatial correlation coefficient of a particular rain rate:

$$\rho(d) \approx \exp(-\lambda_s \sqrt{d}) \tag{15.11}$$

where d, km, is the approximating diameter of the rain extent and $\lambda_s \approx 0.22, \text{km}^{-1/2}$. Hosoya et al. [15] suggest $\lambda_s = 0.25$ based upon their data obtained for the Yokosuka and Yokohama areas of Japan.

Yaozong [14] gives an approximation for the horizontal path distance which is equivalent to the actual distance through a given rain rate:

$$d = d_o R_m^{\lambda_r} \ km \tag{15.12}$$

where

R_m is the median value of rain rate, mm/hr

$d_o =$ $30 + 15[\frac{R_p(.01\%)}{18} - 1]$

$R_p(.01\%)$ is the statistal value of point rain corresponding to 0.01% of the time

λ_r ≈ 0.32, depending upon climate.

Distribution of storms

Storm spacing based upon satellite data over Nigeria, [16] is:

$$S \approx \frac{1.4R^{1.3153}(96.54 - R)}{R - 3.22} \tag{15.13}$$

where R, km, is the radius of the over-cast cloud area as seen by satellite. An average value for R was found to be about 48 km, [16]. Rain areas within an over-cast cloud area varied from 3 to 4 and had an average radius of about 13 km. Goldhirsh and Musiani [17] indicate the median diameter of a rain cell is 2 km, the mean diameter is 2.6 km and the 90 percentile is 5 km for the cell center rain rate range of 2 to 90 mm/hr.

15.2.3   Snow

Kharadly and Choi, [18] use for snow one-way absorption the common approximation that is used for rain attenuation:

$$\alpha \approx a(f)R^{b(f)} \tag{15.14}$$

where R is the water equivalent of rain rate, expressed in mm/hr. From [18] the frequency variable terms for a temperature of 20°C for a frequency range of 8 to 30 GHz are:

$$a(f) = 0.00771F_{GHz}^{1.23}$$
$$b(f) = 1.5 - 0.275/\log F_{GHz}$$

15.2.4   Fog and Clouds

Altshuler [19] gives an approximation for fog attenuation for the frequency range, 10 to 100 GHz, as:

$$A = -1.347 + 0.0372\lambda + \frac{18}{\lambda} - 0.022T \,, \mathrm{dB/km/[g/m^3]} \tag{15.15}$$

where $[\cdot]$ indicates A is expressed in g/m^3 units of water density of fog. The water density for fog can be approximated, knowing visibility, $M(\mathrm{km})$, by [19]:

$M \approx (\frac{a}{V})^{1.54}$, g/m^3

$a \ = 0.024$ for radiation fog or where drop diameters are between 0.3 to 10 microns

$= 0.017$ for very small drop sizes as in the case of a dense haze

$$\tag{15.16}$$

For heavy advection fog where droplet diameters may be as large as 100 microns $M = 0.37$ g/m^3, [39]. This value is in the range of cloud water droplets for which M varies 1 g/m^3 for cumulus clouds to 0.15 g/m^3 for stratus-stratocumulus clouds.

Altshuler and Marr [20] give an approximation, said to be good for 15 to 100 GHz, for cloud attenuation at millimeter wave frequencies as:

$$
\begin{aligned}
A &= \left(-.0242 + 0.00075\lambda_{mm} + 0.403/\lambda_{mm}^{1.15}\right)\left(11.3 + \rho_w\right)D(\theta)\,, \text{dB} \\
D(\theta) &= \operatorname{cosec}(\theta) && \text{for } \theta > 8° \\
&= [(a_e + h_e)^2 - a_e^2 \cos^2\theta]^{0.5} - a_e \sin\theta && \text{for } \theta \le 8°
\end{aligned}
\tag{15.17}
$$

where λ_{mm} is the wavelength in mm, θ is the elevation angle, deg., a_e is the effective earth radius, 8497 km, h_e is the effective height of the attenuating portion of the atmosphere, and ρ is the surface absolute humidity, g/m^3.

$$
h_e = 6.35 - 0.302\rho\,, \text{km} \tag{15.18}
$$

Another approximation is suggested by Liebe et al [22]:

$$
\begin{aligned}
\alpha &= a(f)T^{b(f)}\rho_w\,, \text{dB/km} \\
a(f) &= x_o + x_1 F_{GHz} + x_2 F_{GHz}^2 \\
b(f) &= y_o + y_1 F_{GHz} + y_2 F_{GHz}^2 \\
T &= 300/(T(°C) + 273.15) \\
\rho_w, &\quad \text{see Table 15.1} \\
x_i \text{ and } y_i, &\quad \text{see Table 15.2}
\end{aligned}
\tag{15.19}
$$

The use of the ρ_w term permits the application of the above realtionships for attenuation estimates for clouds as well as fog.

Table 15.1: Constants for second-order approximation to $a(f)$ and $b(f)$ terms for fog and cloud attenuation.

Freq. (GHz)	x_0	x_1	x_2
0 to 100	0	2.18×10^{-3}	3.9×10^{-4}
100 to 1000	-2.24	7.02×10^{-2}	-2.05×10^{-5}
	y_0	y_1	y_2
0 to 250	9.73	-8.92×10^{-2}	1.73×10^{-4}
250 to 1000	-1.12	-3.04×10^{-3}	3.6×10^{-7}

Table 15.2: Typical values of fog and cloud water content, g/m^3.

Light fog[1]	0.001
Moderate fog[1]	0.05
Heavy[1]	1
High cirrus ice clouds[1] $\leq$	0.001
Stratus[2]	0.25
Stratocumulus[2]	0.45
Cirrostratus[2]	0.55
Nimbostratus[2]	0.65
Cumulus[2]	0.75
Heavy cumulus[1]	1
Rain clouds[1]	2

(1) reference [22]
(2) reference [21]

15.2.5 Dust

Ghobrial and Skarief [23] give data that shows the specific attenuation in dust storms is:

$$\alpha = \frac{AF_{GHz}}{V^\gamma}, \text{ dB/km} \tag{15.20}$$

where V is the visibility, km, typically 0.01 to 1. $\gamma \approx 1.07$ for the Sudan, values for A are as shown in Table 15.3:

Table 15.3: Values of constant A for attenuation in dust storms.

Polarization	Dry dust	Dust with 4% water
vertical	2.51×10^{-5}	4.1×10^{-5}
horizontal	4.9×10^{-5}	$8.6 \times 10{-5}$

Abdulla et al. [24] present data shown in Figure 15.5 for falling dust. The upper curve is for 10 m visibility.

15.2.6 Foilage

According to Vogel [52] the one-way loss through road-side trees in the United States can be approximated by:

$$L_F \approx 0.19 F_{GHz}^{0.284} d^{0.59} \tag{15.21}$$

where d is the depth of a tree grove traversed by the propagagtion path, m.

Distribution of the loss through road-side trees, in this case given as fade depth, is shown in Figure 15.6.

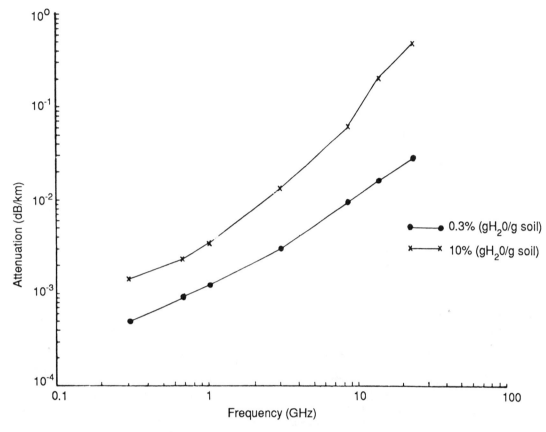

Figure 15.5: Specific attenuation for falling dust, from [24].

Vogel and Goldhirsh [26] summarize the linear-curve data of Figure 15.6 and other similar data with the relationship:

$$L_F = -M \ln P + B, \text{ dB}$$
$$M = 3.44 + 0.0975\theta - 0.002\theta^2$$
$$B = -0.443\theta + 34.76$$

where P is the probability of not exceeding a given L_F and θ, deg, is the elevation angle.

For the rain forests of India, Tewari et al, [28] suggest the use the solid-line curve data in Figure 15.7 for specific attenuation. Note that in this case specific attenuation is in dB/m. They found that for distances greater than 0.4 km, a loss term that is a function of distance must be added to what one would obtain for a distance of 0.4 km using the data of Figure 15.7. These additive loss data are shown in Figure 15.8. The authors surmise that the additive loss is due to a change of the principal mode of propagation from that of penetration for distances less than 0.4 km to a path along the forest-air boundary for distances greater than 0.4 km.

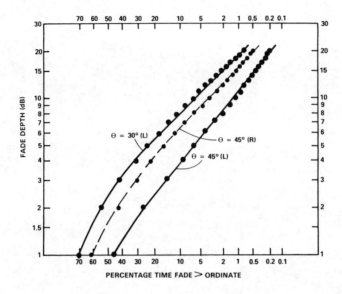

Figure 15.6: Distribution of attenuation at 0.9 GHz through trees along a roadside. The R and L notation denotes right and left side driving of the measuring equipment and therefore reperesent different path lengths through the trees. θ is the elevation angle of the measured path. From [27].

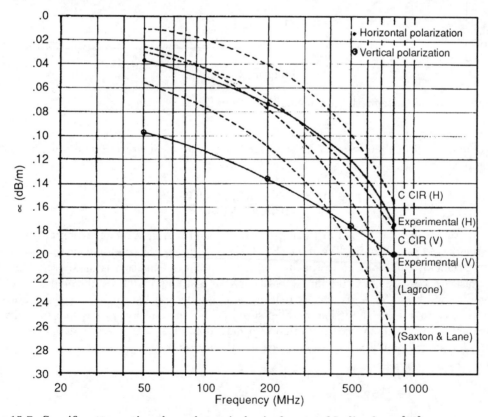

Figure 15.7: Specific attenuation through tropical rain forests of India, from [28].

(a)

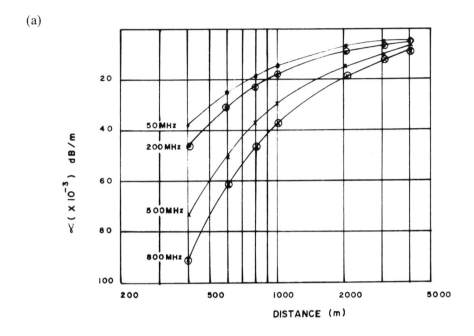

(b)

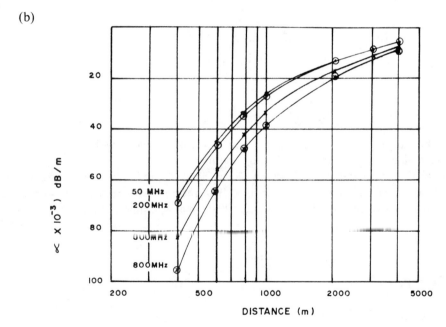

Figure 15.8: Distance dependence of specific attenuation for lateral wave mode; (a) horizontal polarization, (b) vertical polarization, from [28].

15.3   Diffraction

15.3.1   Knife-edge Diffraction

From data presented by Kirby [29] it can be surmised that the one-way propagation factor F past a knife-edge obstacle is:

$$20 \log F \approx -6.65 - 8.19v + 0.98v^2 \,, for - 0.75 \le v \le 3$$
$$0, \quad \text{for } v < -0.75 \tag{15.22}$$
$$v = h\sqrt{(2/\lambda)(1/d_r + 1/d_t)}$$

where h is the distance between the top of the knife-edge obstacle and the direct path between the radar and the target. h is positive when this direct ray passes through the obstacle. It is negative if the direct ray passes over the obstacle. d_r is the distance between the radar and the obstacle and d_t is the distance between the target and the obstacle.

15.3.2   Diffraction of 4/3 Earth

The one-way propagation factor can be approximated by following the method presented by Blake [2], p.273:

$$
\begin{aligned}
F &= V(X)_{dB} + U(Z_1)_{dB} + U(Z_2)_{dB} \\
V(X)_{dB} &= 10.99 + 10 \log X - 17.55X \\
X &= R_{km} F_{GHz}^{1/3}/19.02 \\
U(Z_{1,2})_{dB} &= 20 \log Z_{1,2} \,, for Z_{1,2} \le 0.6 \\
&= -4.3 + 51.04[\log(Z_{1,2}/0.6)]^{1.4} \,, for 0.6 < Z_{1,2} < 1 \\
&= 19.85(Z_{1,2}^{0.47} - 0.9) \,, for Z_{1,2} \ge 1 \\
Z_1 &= h_r F_{GHz}^{2/3}/21.3 \\
Z_2 &= h_t F_{GHz}^{2/3}/21.3
\end{aligned}
\tag{15.23}
$$

where R_{km}, km, is the range to the target, and $h_{r,t}$ are the radar and target heights, m, above ground. Figure 15.9 is an example result using the method.

There is an intermediate region between the regions of predominance of either diffraction or reflection effects in which both effects combine. The intermediate region is in the vicinity of the ray tangent to the earth surface. Barton [25], p. 298 discusses a method for manually interpolating between the worst effects of each. Ayasli [30] discusses a computer method of finding the weighted sum of both effects.

15.4   Refraction

15.4.1   Index of Refractivity

For air containing water vapor the refractivity, N, is [31]:

$$N = 77.6\frac{P}{T} + 3.73 \times 10^5 p/T \tag{15.24}$$

where P is barometric pressure, mbars (1mm Hg = 1.3332 mbars), T is absolute temperature, K, and p is the partial pressure of water vapor, mbars. N is the scaled-up index of refraction n, where $N = (n-1) \times 10^6$. The first term of N can be thought of as the dry atmosphere refractivity, N_{dry}, and the second as the wet atmosphere refractivity, N_{wet}.

N typically varies with altitude and is approximated often with an exponential function [6], p.120:

$$N(h) = N_s \exp(-h/H_N) \tag{15.25}$$

where h is the altitude above sea level, km, and H_N is the scale height, 7 km. N_s, the surface refractivity, varies with atmospheric conditions seasonally and geographically, but is often taken as 313 in the above equation and for which then, the equation is referred to as an exponential reference atmosphere. Costa [32] used other models for N to study atmospheric inhomogeneities. For:

1. Stratification for a horizontally uniform refractive index:

$$N(h) = N_o + k_N h + \frac{\Delta N}{\pi} \arctan \left[12.63 \left(\frac{h - h_o}{\Delta h} \right) \right] \tag{15.26}$$

with for example the following values for the terms used by Costa:

$N_o = 300$
$k_N = -40$, N units/km
$\Delta h = 100$ m
$\Delta N = -20$, N units

2. Inhomogeneity in the horizontal plane, such as for coastal regions:

$$N(h) = N_o + k_N h + \frac{\Delta N}{\pi} \arctan \left[12.63 \left(\frac{h - h_o}{\Delta h} \right) \right] \exp \left[- \left(\frac{x - x_o}{\Delta x} \right)^2 \right] \tag{15.27}$$

with the following example values used by Costa:

$x_o = 25 \text{km}$
$\Delta x = 12.5 \text{km}$

3. For wave-like structures:

$$\begin{aligned} N(h, x) = N(h) = N_o + k_N h + \frac{\Delta N}{\pi} \arctan \left[12.63 \left(\frac{h - h_o}{\Delta h} \right) \right] \\ + \delta N \sin \left[2\pi \left(\frac{h - h_s}{\Delta h_s} \right) \right] \exp \left[- \left(\frac{h - h_e}{\Delta h_e} \right)^2 - \left(\frac{x - x_o}{\Delta x} \right)^2 \right] \end{aligned} \tag{15.28}$$

with the following example values used by Costa:

$\delta N = 1.6$, N units
$\Delta h_s = 20 \text{m}$
$\Delta h_e = 20 \text{m}$
$x_o = 25 \text{km}$
$\Delta x = 10 \text{km}$

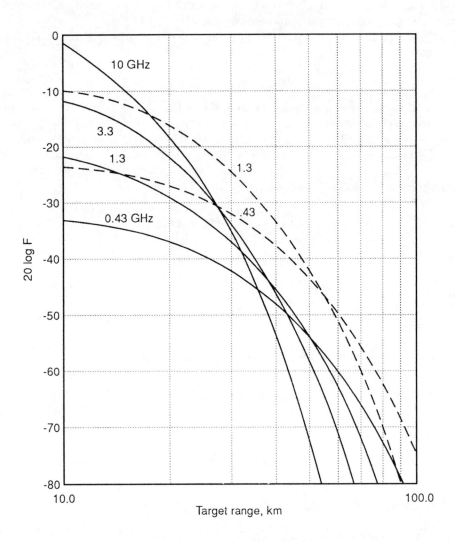

Figure 15.9: One-way propagation factor. Antenna height is 10 m for the solid curves and 30 m for the broken curves. The target height is 3 m.

15.4.2   Effective Earth Radius

Rather than use the more complicated exponential atmosphere model a linear model is used. With a linear model, the effect of refraction can be accounted for in propagation studies with the use of an effective earth radius, $a_e = ka$. k is a constant multiplyer of the actual earth radius, a. k is found from:

$$k = (1 + a\frac{dn}{dh})^{-1} \qquad (15.29)$$

where dn/dh is the change in the index of refraction with altitude. Robertshaw, [53], in his studies of the use of effective earth radius, approximates k as a function of N:

$$k = [1 - 0.04665 \exp(0.005577N)]^{-1} \qquad (15.30)$$

Hanle [35] discusses a statistical variation of k and says:

$\sqrt{k}$ varies ±2% for 50% of the time and terrain situations
±10% for 90%
±30% for 99%

Path clearance criteria

For essentially free space path between low-altitude targets and ground radar the vertical clearance between the ray and the highest obstacle should exceed 0.6 of the first Fresnel zone radius for a given k. The first Fresnel radius is [36]:

$$F_1 = \left(\lambda\frac{d_1 d_2}{d_1 + d_2}\right)^{1/2} \qquad (15.31)$$

where d_1 is the distance between the radar and the highest obstacle between the radar and the target, and d_2 is the distance between that obstacle and the target. The earth bulge should be added to the obstacle height. The bulge is [36]:

$$h_o = \frac{d_1 d_2}{2ka} \qquad (15.32)$$

15.4.3   Standard Range-Height-Angle Charts

Ground radar

Blake [37] has created range-height-angle charts useful for plotting vertical-plane coverage diagrams for earth-based radar systems and for estimating radar target height for a given range and elevation angle. The charts are shown in figures 15.10 through 15.13. The charts are based on the above standard reference atmosphere. Ionospheric refraction is not taken into account, therefore, the chart in 15.13 should be used only at frequencies above about 1 GHz, or at somewhat lower frequencies at night. The range value in the charts is that corresponding to the time of travel from the radar to the target and back. The angle is the direction of the ray at its origin with respect to the local horizontal.

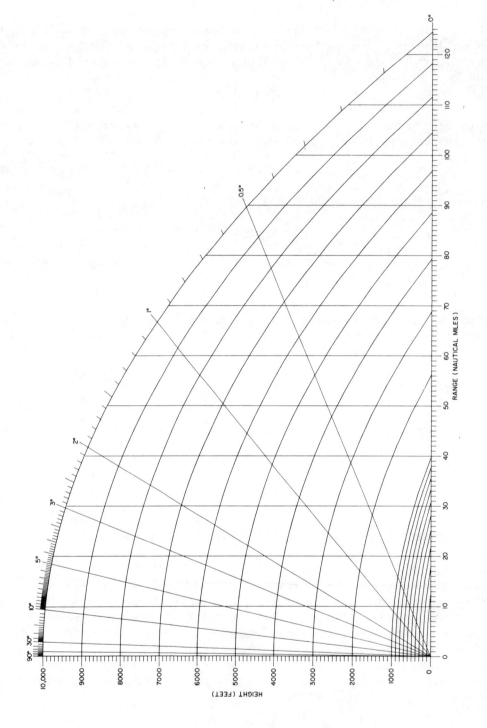

Figure 15.10: Radar range-height-angle chart for exponential atmosphere - low altitudes [h (ft), R (nm)]. From Blake, [37].

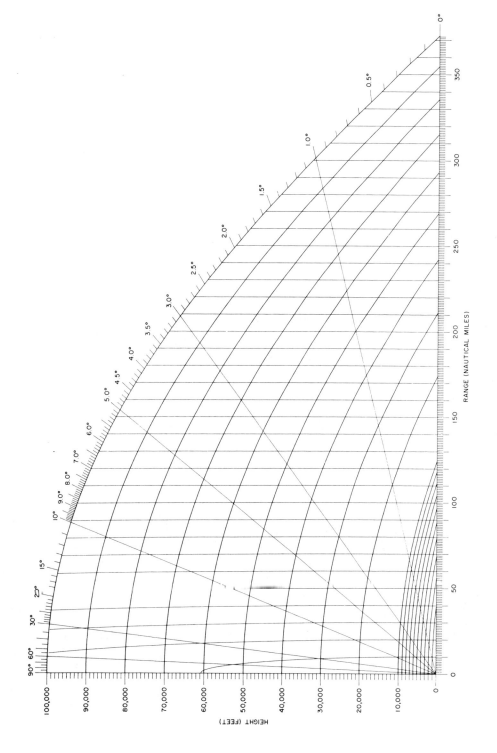

Figure 15.11: Radar range-height-angle chart for exponential atmosphere - high altitudes [h (ft), R (nm)]. From Blake, [37].

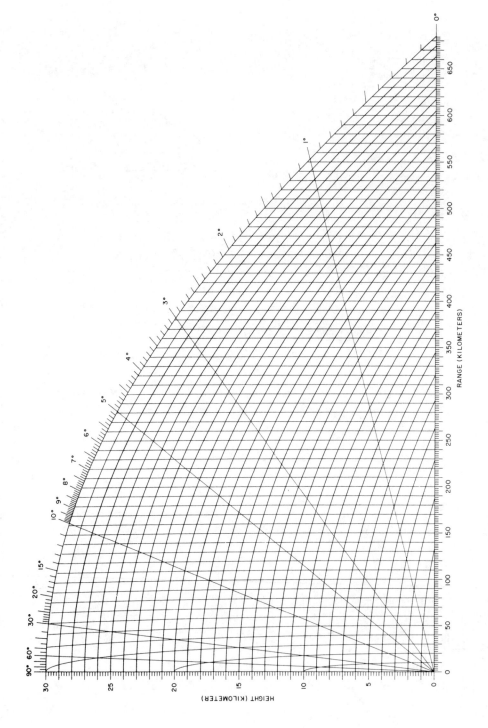

Figure 15.12: Radar range-height-angle chart for exponential atmosphere - high altitudes [h (km), R(km)]. From Blake, [37].

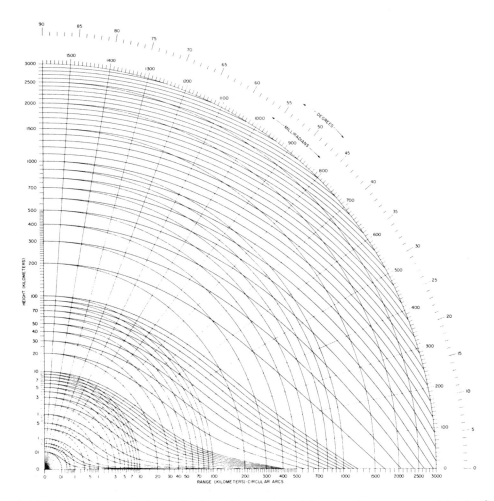

Figure 15.13: **Radar** range-height-angle chart for exponential atmospheres - space altitudes [h(km), R (km)]. From Blake, [37].

Airborne radar

Range-height-angle charts with constant range circle for the standard exponential atmosphere produced by Bauer, [38] are shown in figures 15.14 and 15.15. The limited range of elevation angles does not permit use for the near-range low- and high-altitude situations. However the range of values shown will apply to the majority of applications.

Range-height-angle charts for a more extensive range of elevation angles has been made by Tank [39]. These charts are shown in figures 15.16 through 15.21 for platform altitudes of 10- to 60-kft and a US standard atmosphere. In these charts the outer-range change of slope for the 0 altitude indicates the onset of obscuration of the rays by the earth's horizon.

15.4.4   Range Correction

Ray tracing using the standard exponential atmosphere is a common method used to determine the difference between the radar range and the true target range. Measurement results reported by Robertshaw [53] and [34] were used to produce a simple correction factor to be applied to the value of true target range as determined by use of the standard exponential atmosphere model. Such measurement data were obtained over a three month period of three years (May, June, and July of 1979, 1980, and 1981) for the areas of Hannover and Munich Germany, Dhahran Saudi Arabia, and Pohang South Korea. The general expression for range correction as determined by Robertshaw is:

$$R_c = 0.42 + 0.0319 R_t \sqrt{N_s/H_r} \, , \, \mathrm{m} \tag{15.33}$$

where R_t, km, is the radar reported range of a target 1 km above sea level and H_r, km, is the radar altitude. The target range and radar height conditions for equation 15.33 are: $40 \leq R_t \leq 200$km and $4.5 \leq H_r \leq 20$km. The reference for the correction is determined by ray tracing using the standard exponential atmosphere.

For satellite applications, variations in the electrical path length in the troposhere in excess of free space are, [48]:

$$\Delta R \approx (1.459 + 0.00296 N_s)/\sin \theta_e \, , m \tag{15.34}$$

where N_s is the sea level refractivity and $\theta_e > 10°$ is the ray elevation angle. Additional discussion and reference to another source of range correction is given in Chapter 10.

15.4.5   Ducting

On average, the exponential decrease of refractivity of the atmosphere is about 39 N units/km, [39]. If $N \geq -157N$ units/km a radar ray will be refracted with a ray curvature equal to or greater than earth curvature. Subsequent ray reflection from earth or a bottom layer of a change in refractivity followed by repeated downward bending of the ray will tend to trap the ray propagation; forming a so-called duct. It is common practice to use a modified value for refractivity to indicate the predicted occurance of ducting. The modified refractivity is:

$$M = N + 0.157h \tag{15.35}$$

where h is in meters. Whenever $dN/dh \leq 157N$ units/km, $dM/dh \leq 0$; the indicator of ducting.

An expression used, [40] [41], describing the modified refractivity position is:

$$M(h) = M_0 + 0.125h - 0.125d \ln \left(\frac{h + 1.5 \times 10^{-4}}{1.5 \times 10^{-4}} \right) \tag{15.36}$$

where M_0 is the modified refractivity at the earth surface, h is in meters, and d is the duct thickness. The duct thickness is defined as the height difference between the top of the duct where $dN/dh = -157N$ units/km and the bottom of the duct where either $dN/dh = -157$ or the earth surface occurs.

The maximum trapped wavelength is, [39]:

$$\lambda_{max} = 2.514 \times 10^{-3} [\frac{(N_r - N_{top})}{d} - 0.157)^{1/2} d^{3/2}] \tag{15.37}$$

where N_r is the refractivity at the height of the radar antenna (assumed to be within the duct) and N_{top} is refractivity at the top of the duct.

The probability of an ocean duct height being less than a given value is illustrated in Figure 15.22. From these data one may approximate the frequency of occurrence of a given duct height, for heights greater than 1 m, over the ocean with:

$$p(h) \approx a h^{(b-1)} \exp(-h^b/c) \tag{15.38}$$

where the parameters a, b, and c are given in Table 15.4.

Table 15.4: Parameters used in estimating occurrence of duct height, after [41] and [42].

Water area	a	b	c
Aegean sea	0.0097	2.3	475
North sea	0.053	2.1	70
San Diego	0.029	2.1	150

15.4.6 Lens-effect

The lens loss, [43] [44], for radar altitudes of 0-, 5-, 10-, and 20-kft is shown in Figures 15.23 through 15.26. The data presented are based upon a computer program [44] which had the limitation of working with only positive elevation angles. The results are appropriate for ground-based radars or airborne radars looking upward or airborne radars in which the target is within the horizon. For the airborne radar application in which the target is within the horizon and at an altitude less than the radar one would use the elevation angle existing at the target rather than at the radar. However, for air-to-air beyond-the-horizon applications one must find the minimum ray altitude point. Figures 15.16 through 15.21 will be useful for this purpose. The lens loss is computed for two paths from this minimum ray altitude point: one path to the radar and the other path to the target. Again figures 15.16 through 15.21 will be useful for determining the individual positive elevation angles from the minimum ray altitude point to the radar and the target.

The data in Figure 15.25 and Figure 15.26 were produced with the computer program provided by Shrader and Weil, [44]. The same exponential reference atmosphere model was used for these data as in [44]. The data shown are for a point target. As stated in [44] the lens loss for distributed targets is one-half of that for point targets.

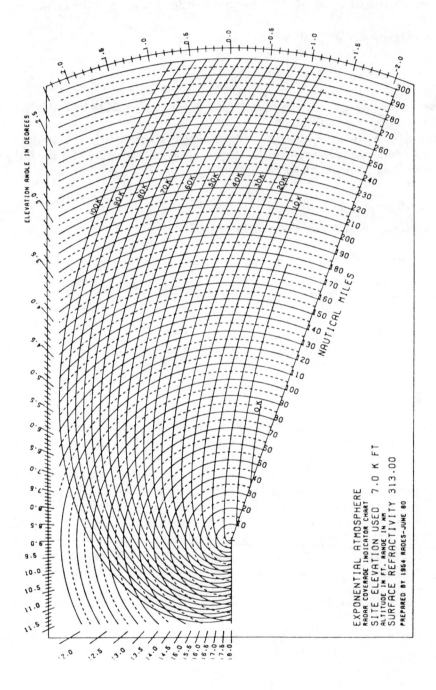

Figure 15.14: Range–height–angle chart for platform altitude of 7 kft. From Bauer [38].

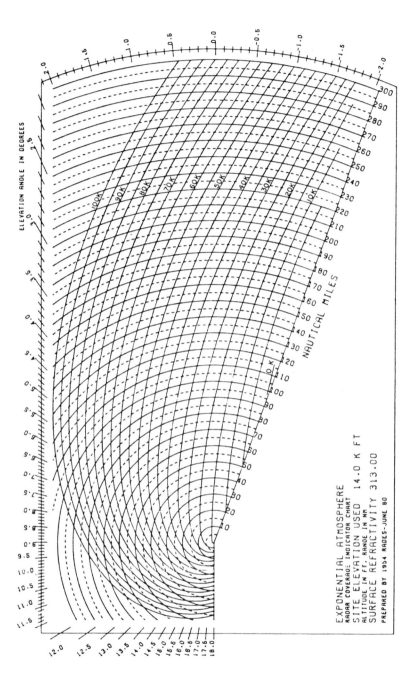

Figure 15.15: Range-height-angle chart for platform altitude of 14 kft. From Bauer [38].

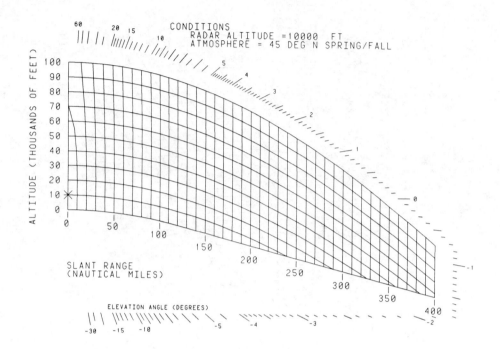

Figure 15.16: Range-height-angle chart for platform altitude of 10 kft. From Tank [39].

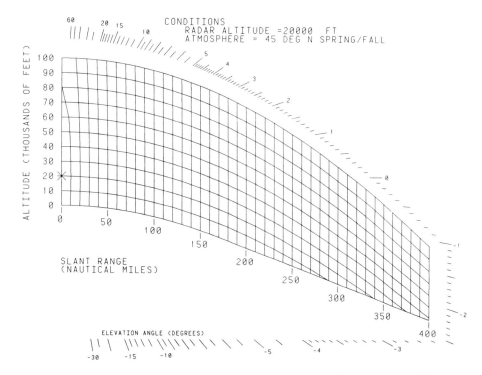

Figure 15.17: Range-height-angle chart for platform altitude of 20 kft. From Tank [39].

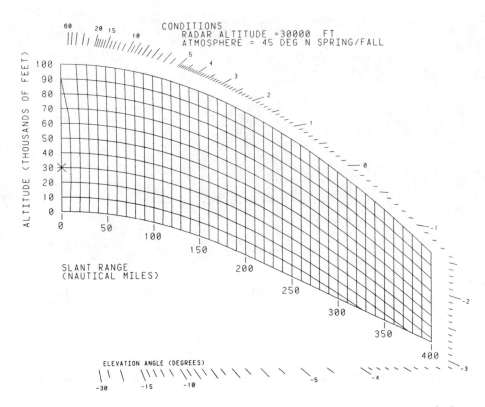

Figure 15.18: Range-height-angle chart for platform altitude of 30 kft. From Tank [39].

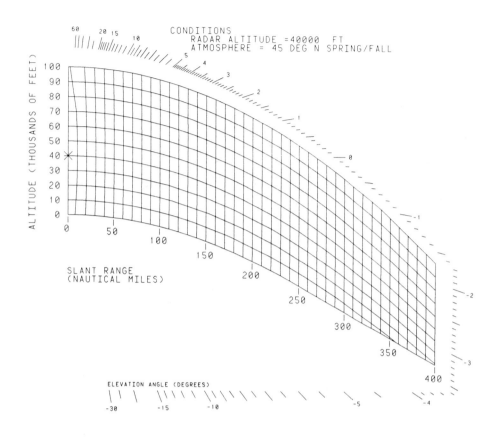

Figure 15.19: **Range-height-angle chart** for platform altitude of 40 kft. From Tank [39].

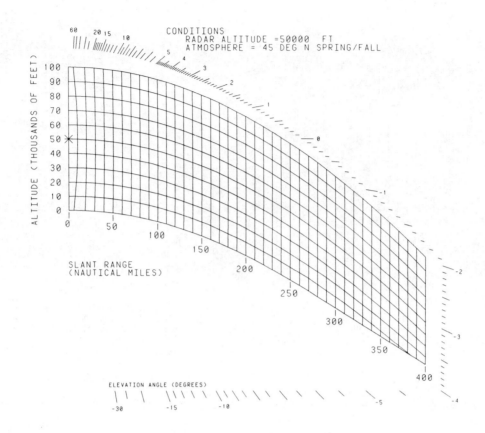

Figure 15.20: Range-height-angle chart for platform altitude of 50 kft. From Tank [39].

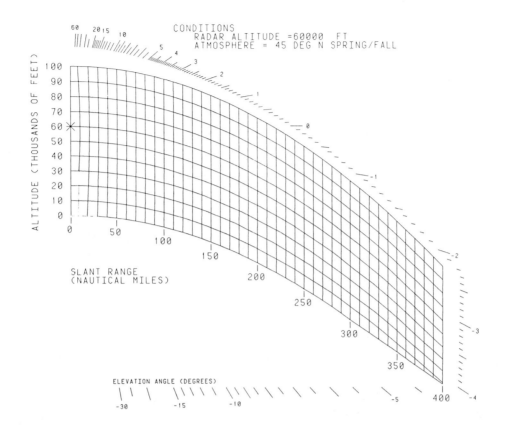

Figure 15.21: Range-height-angle chart for platform altitude of 60 kft. From Tank [39].

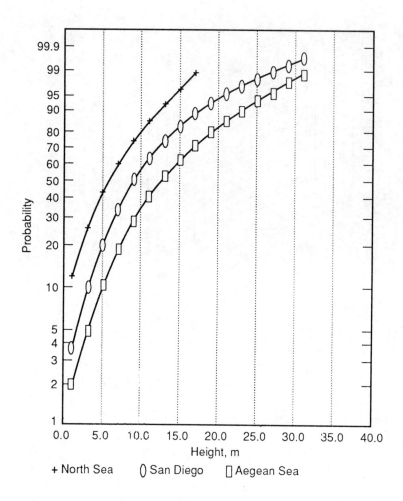

Figure 15.22: Cumulative distribution of evaporation duct heights.

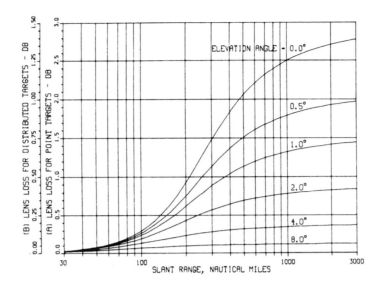

Figure 15.23: Two-way Lens-effect loss for antenna height of 0 kft and sea level refractivity of 313. From [44].

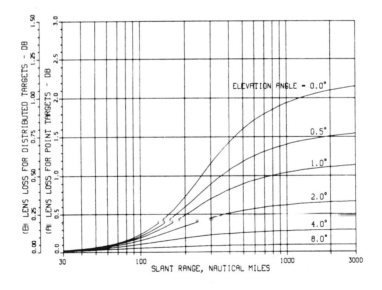

Figure 15.24: Two-way Lens-effect loss for antenna height of 5 kft and sea level refractivity of 313. From [44].

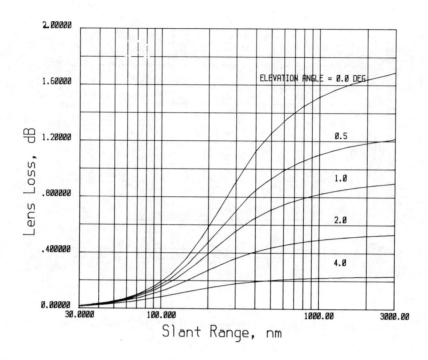

Figure 15.25: Two-way Lens-effect loss for antenna height of 10 kft and sea level refractivity of 313. Based on computer program obtained from [44].

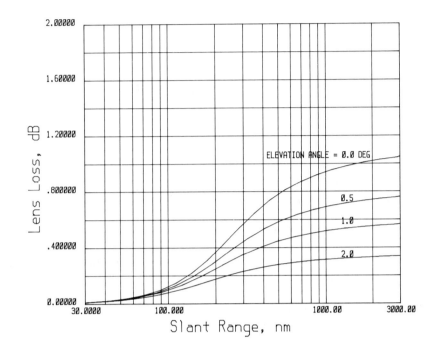

Figure 15.26: Two-way Lens-effect loss for antenna height of 20 kft and sea level refractivity of 313. Based on computer program obtained from [44].

15.5   Scattering

15.5.1   Pattern Factor

For flat earth the approximation for the propagation factor that accounts for multipath is, [45], p.443:

$$F^4 = 16 \sin\left(\frac{2\pi H_r H_t}{\lambda R}\right) \tag{15.39}$$

where H_r is the radar antenna height, H_t is the target height, and R is the range to the target. With $n = (2H_r H_t / \lambda R)$:

$$
\begin{aligned}
F^4 &= 0 \text{ for } n = 0, 1, 2, \ldots \\
&= 16 \text{ for } n = 0.5, 1.5, 2.5, \ldots \\
&= 1 \text{ for } n = 0 + 1/6, 1 - 1/6, 1 + 1/6, 2 - 1/6, 2 + 1/6, \ldots
\end{aligned}
$$

Table 15.5 may be useful for recording the sequence of solutions for the term $n = (2H_r H_t / \lambda R)$. The solution can be for any one parameter; H_r, H_t, λ, or R. For instance one may want to use some value for R and solve for the sequence of radar wavelengths for which one may expect a multipath null, or lobe.

Table 15.5: Form for filling in sequence of solutions for the flat-earth multipath propagation factor.

F^4	0 (-∞,dB)	1 (0 dB)	16 (12 dB)	Parameter values:
n	//////////	//////////	//////////	
1/6	//////////		//////////	$R = $ _____
0.5	//////////	//////////		
1-1/6	//////////		//////////	$H_r = $ _____
1		//////////	//////////	
1+1/6	//////////		//////////	$H_t = $ _____
1.5	//////////	//////////		
2-1/6	//////////		//////////	$\lambda = $ _____
2		//////////	//////////	
2+1/6	//////////		//////////	
2.5	//////////	//////////		
3-1/6	//////////		//////////	
3		//////////	//////////	
3+1/6	//////////		//////////	
3.5	//////////	//////////		
4-1/6	//////////		//////////	
4		//////////	//////////	
4+1/6	//////////		//////////	
4.5	//////////	//////////		

15.5.2   Path Length Difference

From Kerr [46], the sequence of the following equations can be used to find for a 4/3 earth an approximation for the path length difference, δ_r, between the direct and reflected rays:

$$
\begin{aligned}
p &= \tfrac{2}{\sqrt{3}}\left[a_e(H_1 + H_2) + (R/2)^2\right]^{1/2} \\
\phi &= \arccos\left[2a_e(H_1 - H_2)R/p^3\right] \\
H_1 &\geq H_2 \\
R_1 &= R/2 + p\cos[(\phi + \pi)/3] \\
R_2 &= R - R_1 \\
S_1 &= R_1/\sqrt{2a_e H_1} \\
S_2 &= R_2/\sqrt{2a_e H_2} \\
\delta_r &= (2H_1 H_2/R)(1 - S_1^2)(1 - S_2^2)
\end{aligned}
\tag{15.40}
$$

where H_i are the target and radar heights, R is the radar slant range, and a_e is the 4/3 earth radius.

15.5.3   Frequency Shift Between Direct and Multipath

The frequency shift between the direct and multipath return is, [47]:

$$
f_{Dm} = \frac{2V}{\lambda}\frac{H_r H_t}{R^2}
\tag{15.41}
$$

15.6   Fluctuation

15.6.1   Amplitude Fluctuation

Fluctuation of a signal that propagates through the atmosphere occurs due to disturbance of the refractive index. Karasawa et al. [49] have shown the relationship at 11.5 GHz is:

$$
\sigma_x \approx 0.145 + 0.00525 N_{\text{wet}}\,,\text{dB}
\tag{15.42}
$$

where σ_x is the monthly average standard deviation of one-way signal scintillation and that the hourly fluctuation is log-normal distributed. For N_{wet} see 15.4.1. σ_x is proportional to $f^{0.45}L^{1.3}$, where f is the operating frequency and L is the path length.

Filho et al. [50] conclude that for frequencies where there is substantial absorption, 36-and 55-GHz, the fluctuation distribution is not the usual log-normal distributed but a product of two distributions; one for scattering (ρ_s), the other for absorbtion (ρ_a):

$$
\begin{aligned}
\rho_s(2x_s) &= \frac{1}{\sqrt{2\pi}\sigma_{2x_s}}\exp\left(\frac{-x_s^2}{2\sigma_{2x_s}}\right) \\
\rho_a(2x_a) &= \frac{\exp(2x_a)}{a^2}\exp\left[\frac{-(1 + \exp(2x_a))}{a^2}\right]I_o\left(\frac{2\sqrt{\exp(2x_a)}}{a^2}\right)
\end{aligned}
\tag{15.43}
$$

a is a distribution shape parameter. They further note that $\sigma_x = 0.03dB$ at 36 GHz and .061 at 55 GHz.

15.7   Ionosphere

15.7.1   Correlation Time Through the Ionosphere

From an uncited reference the frequency, amplitude and minimum correlation time for propagation through the ionosphere is summarized in Table 15.6. There are two sets of data for the frequency 254 MHz.

Table 15.6: Ionospheric effects on frequency and amplitude fluctuations and correlation time of signals.

Satellite height	Radar frequency (GHz)	Fluctuation frequency (Hz)	Amplitude fluctuation (dB)	Correlation time (s)
Geostationary	.254	0.15	1.1	1.13
"	.254	0.2	5.5	0.85
"	4	0.14	1.2	
1100 km	.15	1	1.27	0.17
1100 km	.4	1.9	5.5	0.09

15.7.2   Faraday Loss and Rotation Angle

Faraday loss, L_F, and rotation angle, θ_R, for linear polarization is [51]:

$$L_F = 20 \log \mid \cos \theta_R \mid$$
$$\theta_R = 81\pi \times 10^{-18} N_T f_H \cos \theta / c F_{GHz}^2 \text{ , radians}$$

(15.44)

where:

f_H is earth's gyromagnetic frequency, Hz

$f_H = 2.84 \times 10^6 B$

B is earth magnetic field, gauss; 239 to 682, see figure 6 of[51].

θ is the angle between earth magnetic field and direction of propagation.

N_T is the number of electrons in a 1 m^2 column along the one-way path from radar to target.

$N_T = N_v / \sin \Phi$

N_V is the number of electrons in a vertical column through ionosphere having 1 m^2 cross section.

N_V nominally varies from 5- to 60 $\times 10^{16}$ with time of day for high sunspot activity and from 2- to 30×10^{16} for low sunspot activity. See [51].

Φ is penetration angle of line-of-sight relative to ionosphere peak altitude.

Faraday rotation angles measured, [52], over a one-year period are shown in Figure 15.27.

c is the velocity of propagation of light, 3×10^8 m/sec.

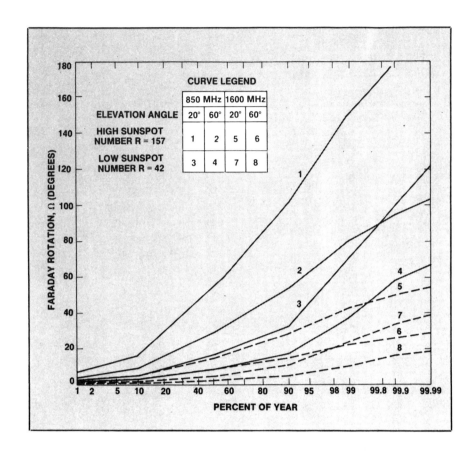

Figure 15.27: Observed Faraday rotation for transmisions from a midlatitude station to a geostationary satellite, from Vogel and Smith [52].

Bibliography

[1] Rose, C., M.J. Gans, "A dielectric-free superconducting coaxial cable," IEEE Trans. MTT-38, no. 2, February 1990, pp. 166–177.

[2] L.V. Blake, *Radar Range-Performance Analysis*, Artech House, Norwood, MA, 1986, p. 208.

[3] L.V. Blake, "Radar attenuation by atmospheric oxygen" URSI-IRE Comm.2, Tropospheric Radio Propagation Session, May 6, 1959.

[4] L.V.Blake, "A guide to basic pulse-radar maximum range calculation" Part 1, NRL 5868, December 28, 1962.

[5] L.V. Blake, personal communication, May 14, 1990.

[6] W.C. Morchin, *Airborne Early Warning Radar*, Artech House, Norwood MA, 1990, figure 3.6.

[7] Altshuler, E., R.A. Marr "A comparison of experimental and theoretical values of atmospheric absorption at the longer millimeter wavelengths," IEEE Trans. AP-36, no. 10, October 1988, pp. 1471–1480.

[8] Galm, J.M., F.L. Merat, P.C. Claspy "Estimates of atmospheric attenuation sensitivity with respect to absolute humidity at 337 GHz," IEEE Trans. AP-38, no. 7, July 1990, pp. 982–986.

[9] Hovanessian, S.A., "Microwave, millimeter-wave and electro-optical remote sensor systems," Microwave Journal, State-ot-the-art Reference, September 1990, pp. 109–129.

[10] Tank, W.G. "Atmospheric Effects," Ch.3, *Airborne Early Warning Radar*, W. Morchin, ed., Artech House, Norwood, MA, 1990, pp. 91–97.

[11] Olsen, R.L., D.V. Rogers, D.B. Hodge, "The aR^b relation in the calculation of rain attenuation," IEEE Trans. AP-26, no. 2, March 1978, pp. 318–329.

[12] Manabe, T., T. Ihara, J. Awaka, Y. Furuhama, "The relationship of raindrop-size distribution to attenuation experienced at 50, 80, 140, and 240 GHz," IEEE Trans. AP-35, no. 11, November 1987, pp. 1326–1330.

[13] Sekine, M., C-D Chen, T. Musha, "Rain attenuation from log-normal and Weibull raindrop-size distributions," IEEE Trans. AP-35, no. 3, March 1987, pp. 358– 359.

[14] Yaozong, F., "Statistical prediction of rain attenuation," IEEE AP-S International Symposium, 1986, June 8-13, 1986.

331

[15] Hosoya, Y., et al, Part V., "Propagation Characteristics of Japan's CS (Sakura) Communications Satellite Experiments," IEEE Trans. AES-22, no. 3, May 1986, pp. 255–263.

[16] Wheeler, N.S., "Aerostat availability as limited by thunderstorms," TCOM Co., ED-TM-110, September 17, 1884.

[17] Goldhirsh, Musiani, "Rain cell size derived from radar," IEEE Trans. Geoscience & Remote Sensing, November 1986.

[18] Kharadly, M.M., A. Choi, "A simplified approach to the evaluation of EMW propagation characteristics in rain and melting snow," IEEE Trans. AP-36, no. 2, February 1988, pp. 282–296.

[19] Altshuler, E.E., "A simple expression for estimating attenuation by fog at millimeter wavelengths," IEEE Tran AP-32, no. 7, July 1984, pp. 757–758.

[20] Altshuler, E.E., R.A. Marr, "Cloud attenuation at millimeter wavelengths," IEEE Trans. AP-37, no. 11, November 1989, pp. 1473–1479.

[21] Cantafio, L., E. Kovaleik, "Space-based radar antennas for weather observation missions," Microwave Journal, January 1990, pp. 131–147.

[22] Liebe, H.J., T. Manabe, G.A. Hufford, "Millimeter-wave attenuation and delay rates due to fog/cloud conditions," IEEE Trans. AP-37, no. 12 , December 1989, pp. 1617–1623.

[23] Ghobrial, S.I., S.M. Sharief "Microwave attenuation and cross-polarization in dust storms," IEEE Trans. AP-35, no. 4, April 1987, pp. 418–425.

[24] Abdulla, S.A., H.M. Al-Rizzo, M.M. Cyril, "Particle-size distribution of Iraqi sand and dust storms and their infuence on microwave communication systems," IEEE Trans. AP-36, no. 1, January 1988, pp. 114–126.

[25] Barton, D.K., *Modern Radar System Analysis*, Artech House, Norwood MA, 1988.

[26] Vogel, W.J., J. Goldhirsh, "Mobile satellite system propagation measurements at L-Band using MARECS-B2," IEEE Trans. AP-38, no. 2, February 1990, pp. 259– 264.

[27] Goldhirsh, J., W.J. Vogel, "Roadside tree attenuation measurements at UHF for land mobile satelitte systems," IEEE Trans. AP-35, no. 5, May 1987, pp. 589–596.

[28] Tewari, R.K., S. Swarup, M.N. Roy, "Radio wave propagation through rain forests of India," IEEE Trans. AP-38, no. 4, April 1990, pp. 433–449.

[29] Kirby, R.C., "Radio-wave propagation," secs. 18.65 - 18.113 in D.G. Fink and D. Christiansen, eds., *Electronic Engineers' Handbook*, McGraw-Hill, New York, 1982.

[30] Ayasli, S., "SEKE: A computer model for low altitude radar propagation over irregular terrain," IEEE Trans. AP-34, no. 8, August 1986, pp. 1013–1023.

[31] Litva, J., "Early results: very low-level propagation measurements on the Ottawa river," IEEE Trans. AP-35, no. 4, April 1987, pp. 469–473.

[32] Costa, E., "The effect of ground-reflected rays and atmospheric inhomogeneities on multipath fading," IEEE Trans. AP-39, no. 6, June 1991, pp. 740–745.

[33] Robertshaw, G.A., "Effective earth radius for refractioin of radio waves at altitudes above 1 km," IEEE Trans. AP-34, no. 9, September 1986.

[34] Robertshaw, G.A., "Range corrections for airborne radar – a Joint STARS study," The Mitre Corp., project 6460, Air Force Electronic Systems Division, ESD-TR-84-169, May 1984.

[35] Hanle, E., "Some new aspects on low-elevation radar coverage," IEEE Int. Radar Conf., 1985, Arlington, VA., May 6-9, 1985, pp. 163-168.

[36] Harvey, R.A., "A subrefractive fading model for microwave paths using surface synoptic meteorological data," IEEE Trans. AP-35, no. 7, July 1987, pp. 832-844.

[37] Blake, L.V., "Radio ray (radar) range-height-angle charts," NRL report 6650, January 22, 1968.

[38] Bauer, K.W., "Range-height-angle charts with lookdown capability," Microwave Journal, October 1981, pp. 89-92.

[39] Tank, W.G., "Atmospheric Effects," *Airborne Early Warning Radar*, W. Morchin, ed., Artech House, Norwood, MA, 1990.

[40] Paulus, R.A., "Evaporation duct effects on sea clutter," IEEE Trans. AP-38, no. 11, November 1990, pp. 1765-1771.

[41] Anderson, K.D., "94-GHz propagation in the evaporation duct," IEEE Trans. AP-38, no. 5, May 1990, pp. 746-753.

[42] Hitney, H.V., R. Vieth, "Statistical assessment of evaporation duct propagation," IEEE Trans. AP-38, no. 6, June 1990, pp. 794-799.

[43] Weil, T.A., "Atmospheric lens-effect; Another loss for the radar range equation," IEEE Trans. AES-9, no. 1, January 1973, pp. 51-54.

[44] Shrader, W.W., T.A. Weil, "Lens-effect loss for distributed targets," IEEE Trans. AES-23, no. 4, July 1987, pp. 594-595.

[45] Skolnik, M.I., *Introduction to Radar Systems*, McGraw-Hill, New York, 1980.

[46] Kerr, D.E., *Propagation of Short Radio Waves*, M.I.T. Radiation Laboratory Series, vol. 13, McGraw-Hill, New York, 1951.

[47] Proceedings of International Radar Conference, Arlington, VA, 1985.

[48] Brussaard, G., D.V. Rogers, "Propagation considerations in satellite communication systems," IEEE Proc. vol. 78, no. 7, July 1990, pp. 1275-1282.

[49] Karasawa, Y., K. Yasukawa, M. Yamada, "Tropospheric scintillation in the 14/11-GHz bands on earth-space paths with low elevation angles," IEEE Trans. AP-36, no. 4, April 1988, pp. 563-569.

[50] Filho, F.C.M., F.H.R. da Silva Mellow, R.S. Cite, "A new model for millimeter wave amplitude fluctuation distribution in an absorbtion band," IEEE AP-S Int. Symp. 1986, June 8-13, 1986.

[51] Brookner, E., W.M. Hall, R.H. Westlake, "Faraday loss for L-Band radar and communication systems," IEEE Trans. AES-21, no. 4, July 1985, pp. 459-469.

[52] Vogel, W.J., E.K. Smith, "Propagation considerations in land mobile satellite transmission," Microwave Journal, October 1985, pp. 111–130.

[53] Robertshaw, G., "How accurate is range correction?," Microwaves & RF, March 1986, pp. 129–132.

Notes

Notes

Chapter 16

Radar Equation

16.1 Introduction

It hardly seems necessary to devote a chapter to the radar equation in retrospect to all the books and computer programs devoted to the subject. However, each time the radar cumulative index, [1], is updated new references are added–showing a continuing advancement in the techniques for design synthesis and prediction of radar performance. We hope to add to these sources of information with the material in this chapter even though we repeat some forms of the radar equation for completeness.

For additional information the reader is referred to the books: [2], [3], and [4], and to the computer programs: [5], [6], [7], [8], [9], [10], and [11].

16.2 General Equation

The general monostatic radar range equation in itself has many forms. The one we use is:

$$(S/N)_1 = \frac{P_k \tau G^2 \lambda^2 \sigma}{(4\pi)^3 k T_o F_n L R^4} \tag{16.1}$$

where

P_k is transmitter peak power.

G is antenna gain. This assumes the transmit antenna gain = receive antenna gain. If such is not the case then G_t and G_r are used.

$(S/N)_1$ is the single-pulse signal-to-noise ratio.

σ is the target monostatic radars cross section.

k is Boltzmann's constant; 1.38×10^{-23}.

T_o is a reference temperature for NF; 290 K.

τ is the transmitted pulse width.

Fn is the noise figure (see Chapter 21).

L is the total set of losses; dissipative, propagation and signal processing.

R is the target distance from the antenna.

16.2.1 Search

Many radars are used for search, therfore we present an adaption of the radar equation for search:

$$(S/N)_i = \frac{P_{avg} A_e T_r \sigma}{(4\pi)kT_o F_n L R^4 \Omega} \tag{16.2}$$

where
$(S/N)_i$ is the integrated signal-to-noise ratio.

P_{avg} is the transmitter average power.

A_e is the effective aperature of the antenna.

T_r is the time to search an angular area, Ω.

The remaining terms are as defined for the basic radar equation. We show the equation solution for (S/N) rather than R to avoid interations for solutions of R when other terms are range dependent. Chapter 5 will be of assistance in determining if (S/N) is adequate for detection. Chapter 18 presents the search equation in a form suitable for specifying a cumulative probability of detection.

16.2.2 Track

$$(S/N)_i = \frac{P_{avg} A_e^2 T_{int} \sigma}{(4\pi)kT_o F_n L R^4 \lambda^2} \tag{16.3}$$

where T_{int} is the radar integration time of a target in track. If a total of N_t targets are being tracked, then the data interval on any one target is $N_t T_{int}$, assuming T_{int} is the same on all targets. If T_{int} is not the same, one will need to construct a time line of the various values of T_{int} to determine the data rate for each target.

16.2.3 Speciality Applications

1. Pulse Doppler, [12]

   Rather than P_k, P_{avg} is used and τ in the numerator is replaced by the bandwidth of a Doppler filter, B_n, in the denominator in the general radar equation. Long et al. [12] provide a similar equation that accounts for transmitter and receiver duty factors.

2. CW

   The radar equation is the same as that for the pulse Doppler radar. B_n may be determined however by the time on the target due to, for example, antenna scanning.

3. Bistatic (see Chapter 13)

4. Cumulative Detection (see Chapter 18)

5. Radar in jamming

6. Radar in clutter

7. Synthetic aperture radar (see Chapter 22)

8. Meteorological Radar (see Chapter 11)

9. Over-the-horizon Radar (see Chapter 14)

10. Laser Radar (see Chapter 9)

11. Others: ground probing, harmonic target, impulse, and random signal.

16.2.4   Field Strength and Incident Power Density

The field strength, without losses, at a distance, R, meters, is:

$$E_f = \sqrt{30PG}/R, \qquad \text{v/m} \tag{16.4}$$

where P, watts is the power delivered to an antenna with gain G. The power density is:

$$\beta = PG/4\pi R = E_f^2/377 \tag{16.5}$$

16.3   Tabular Solutions

Tabular solutions may seem useless in this day of spreadsheets and the abdundancy of specialized computer programs available to solve the radar equation. However, there is always a need to summarize thoughts and show results on a piece of paper.

We offer the tabular worksheet shown in Tables 16.1 and 16.2 with which to work through a top-level design concept for a search radar in a noise only environment. The purpose is to arrive at a nearly congruent set of parameters describing a radar design. We say "nearly congruent" because approximations are used and the chart does not cover the details one would derive with a deeper synthesis using specialized computer programs to predict the effects of clutter and other sources of interference and the unique impact specific waveform selections will have on the parameters. The chart may also serve as an outline for a computer spreadsheet program.

The chart is used to first solve the basic search equation for power-aperture product followed by a solution for the optimum antenna size. After choosing the antenna relative sidelobe level and selecting antenna width and height dimensions the antenna beamwidth are determined. This particular order assumes limitations on antenna width or height are greater than what the beamwidths may impose. The inverse is possible. The weight of the antenna and transmitter is found after setting weight factors for the antenna and the transmitter. The values thus far set and derived are used to find the single-pulse signal-to-noise ratio in Table 16.2 and subsequently the number of coherent pulses integrated and the number of noncoherent intervals based on assumed target acceleration. The technique ends with prediction of the probability of detection. It would normally be expected to require several passes through the chart with which to derive a satisfactory set of parameters.

Table 16.1: Search radar concept development worksheet.

Solution of Radar Search Equation					
Radar Parameter	**Equation Term**	**Numeric Value**	**Numer. (dB)**	**Denom. (dB)**	**Notes and supporting values or equations**
Integrated S/N	$(S/N)_i$			////////	
	$4\pi k T_o$		-193	////////	
System Loss	L			////////	
Noise Figure	F_n			////////	
Target Range, m	R^4			////////	
Angular Vol, sr	Ω			////////	sr. $= 3282.8^o$
Search Time, s.	T_r		////////		
Target RCS, m²	σ		////////		
Power Aperture	PA_e	=	–	=	dBW

From Table 16.1: $PA_e = 10^{PA_e/10} =$ _____ Wm^2

    Set power weight factor, $k_p =$ _____ kg/W

    Set antenna weight factor, $k_a =$ _____ kg/m^2

    Set peak antenna sidelobe level relative to the mainbeam,

$g_{sl} =$ _____ dB

Aperture efficiency, $\eta \approx 1.1 - 0.013 g_{sl} =$ _____;

Aperture Loss $= 10 \log \eta =$ _____, dB.

Included above in system losses.

Antenna Size, $A_r = (k_p PA_e/k_a \eta)^{1/2} =$ _____,

    for equal transmitter and antenna weights.

    Set $A_r =$ _____m^2

Antenna beam broadening factor, $K_3 \approx 0.72 - 0.0134 g_{sl} =$ _____

    Set antenna width, $W =$ _____, m;

        or Azimith beamwidth $\theta =$_____deg.

    Set antenna height, $H =$ _____, m;

        or Elevation beamwidth $\phi =$_____deg.

Azimuth beamwidth $= K_3 \lambda/W =$ _____ r;

    or Antenna width $W = \frac{K_3 \lambda}{\theta} =$_____, m.

    $\times 57.3 =$ _____ deg.

Elevation beamwidth $= K_3 \lambda/H =$ _____ r;

    or Antenna Height, $H = \frac{K_3 \lambda}{\phi} =$_____deg.

    $\times 57.3 =$_____ deg.

Antenna and Transmitter Weight, $k_p P + k_a A_r =$_____ kg.

Table 16.2: Single-pulse solution worksheet.

Single-Pulse (S/N) Solution					
Radar Parameter	**Equation Term**	**Numeric Value**	**Numer. (dB)**	**Denom. (dB)**	**Notes and supporting values or equations**
Peak Power, W	P_k			//////// ////////	$P_k = P/(\tau PRF) =$
Transmitted pulse, s	τ			////////	
Wavelength, m	λ			//////// ////////	
Antenna Gain	$G_t G_r$			//////// ////////	$G = 4\pi/\theta_{AZ}\theta_{EL} = 4\pi A\lambda^2 =$
Target RCS, m^2	σ			//////// ////////	
constants	$kT_o(4\pi)^3$		////////	-171	
Noise Figure	F_n		////////		
System Loss	L		////////		
Target Range, m	R		////////		
One Pulse Signal-to-Noise	$(S/N)_1$	=		–	= dB

Observation Time, T_{ob}

(Azimuth only scan) $= \theta^o_{AZ}/6R\text{pm} = $ _____ , s

(Azimuth and Elevation scan) $= T_r\theta_{AZ}\theta_{EL}/\Omega = $ _____ s.

Set target acceleration, $g = $ _____ ,

then $a_t = 9.81g = $ _____ m/s^2

Limit on coherent processing time, $t_c \approx \lambda/2\sqrt{a_t} = $ _____ , s.

Number of coherent intervals possible, $N_{nc} = T_{ob}/t_c = $ _____ ;

Set $N_{nc} = $ _____

Coherent processing time for one coherent interval possible, $T_c = T_{ob}/N_{nc} = $ _____ ;

Set $T_c = $ _____ , s.

Set the PRF $= $ _____ , Hz

Number of Coherently Integrated Pulses, for one coherent interval, $N_c = T_c\text{PRF} = $ _____

$(S/N)_c$ after coherent integration $= N_c(S/N) = $ _____

Effective number of pulses after noncoherent integration, [19], eq. 5.9.

$N_e \approx N_{nc}^{\exp[-0.04-0.05\ln(N_{nc})]} = $ _____

$(S/N)_i$ after noncoherent integration $= N_e\,(S/N)_c = $ _____

Set probability of false alarm, $P_{fa} = $ _____

If noncoherent integration is used,

Swerling 1, $P_d = P_{fa}^{(1+2\sqrt{N_e}(S/N)_c)^{-1}} = $ _____

If coherent integration only is used,

Swerling 1, $P_d = P_{fa}^{(1+(S/N)_c)^{-1}} = $ _____

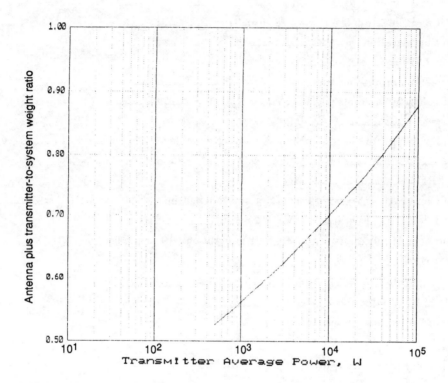

Figure 16.1: Radar system weight dependence on transmitter power and transmitter and antenna weight.

An empirical relationship between airborne radar system weight and antenna and transmitter weight is shown in Figure 16.1. We see that for the lower power radars the antenna and transmitter weight are about one-half of the radar system weight, whereas for the high-power systems the order of 90% of the system weight is due to the transmitter and antenna.

16.4   Parameter Values

Sources for information concerning the selection of values for the radar equation are contained in the various chapters of this book. However, the subject of losses, although covered in these other chapters too, needs further discussion. Radar performance prediction is dependent on the art of selecting the appropriate combination of losses and their associated values. Radar system lossses are a measure of the equipment and propagation inefficiencies and deviations from the performance measures associated with the radar equipment. We hope to quide the reader in selection of a loss budget by showing example budgets. We follow this with approximations for losses not found elsewhere in this book.

16.4.1 Sample Loss Budgets

Table 16.3 illustrates a sample loss budget for an S-band airport surveillance radar application with detection ranges of 40 to 70 km, [14] and Table 16.4 illustrates the losses attributed to a C-band airborne early warning radar, [13]. A example of a loss budget for an airborne air-to-air target track radar is shown in Table 16.5.

Table 16.3: Loss factors for low-PRF systems.

| Source of loss | Loss, dB Type of system | | | Type of loss |
	MTI	MTD	Frequency agile MTI	
Transmission Loss	2.0	2.0	1.0	Dissipative
Atmospheric Loss	1.0	1.0	1.0	Propagation
Beamshape Loss	1.3	1.3	1.3	Signal proc.
Mismatch Loss	1.0	1.0	1.0	Signal proc.
Processing Loss	6.0	3.0	8.0	Signal proc.
Integration Loss	3.2	0.4	1.6	Signal proc.
Fluctuation Loss	8.4	8.4	1.1	Signal proc.
Total	22.9	17.1	15.0	

Table 16.4: Example of C-band losses for an airborne early warning radar, from [13].

Source of loss	Loss, dB	Type of loss
Propagation	6.3	Propagation
Radome	0.8	Dissipative
Beam shape	1.6	Signal processing
Waveguide attenuation	1.2	Dissipative
Transmitter components	1.3	Dissipative
Receiver components	1.9	Dissipative
Clutter filling	0.6	Signal processing
Eclipsing	0.2	Signal processing
IF mismatch	0.4	Signal processing
Straddle	1.0	Signal processing
CFAR	0.7	Signal processing
Processing	2.0	Signal processing
Field degradation	2.0	Signal processing, dissipative and reduced power or gain
Total	20.0	

Table 16.5: Example loss budget for an X-band air-to-air track radar for ranges to 185 km using a non-gimbaled active phased array antenna.

Source of loss	Loss, dB	Type of loss
Doppler weighting	1.4	Signal processing
Range gate straddle	1.0	Signal processing
CFAR	0.7	Signal processing
IF mismatch	1.2	Signal processing
Component	0.2	Dissipative
Antenna element	2.0	Dissipative
Antenna scan	$30\log\cos\theta$	Loss component perf.
Atmospheric	6.0	Propagation
Cloud attenuation	0.8	Propagation
Rain attenuation	5.5	Propagation
Lens effect	0.2	Propagation
Eclipsing	0.6	Signal processing
I and Q recombination	0.2	Signal processing
Clutter settling time	2.0	Signal processing
Detection 2 out of 3 PRFs	1.8	Signal processing
Total	$23.6 + 30\log\cos\theta$	

A sample plot of losses as a function of frequency is shown in Figure 16.2, from what is now an anonymous study. The curve noted as signal processing loss also includes dissipative losses in the LNA, T/R switches, receiver protector as well as VSWR mismatch losses. The rain curve is for 4 mm/hr over an extensive area. This particular rain loss curve is approximated by:

$$\alpha \approx aF_{GHz}^{b}, dB \tag{16.6}$$

where $a = 0.00282$ and $b = 3.9$.

The dissipative loss and signal processing loss curve is approximated by a function similar to the above except that $a = 7.7$ and $b = 0.2$. Other data suggested that dissipative only losses followed the above relationship with $a = 1.2$ and $b = 0.2$, thus indicating fixed signal processing losses for the frequency range of the plotted data in Figure 16.2.

In addition to losses one may want to use in studies directed to frequency selection, equations and values for noise figure in Chapter 21 or an approximation:

$$\begin{aligned} NF &\approx a + bF_{GHz}^{c}, \text{ dB} \\ a &\approx 0.5 \\ b &\approx 1 \\ c &\approx 0.45 \end{aligned} \tag{16.7}$$

16.4.2 Losses Not Found in Other Chapters

Radome loss

Kim and Sandberg, [15], use the following approximation for a ground based X-band communications system radome.

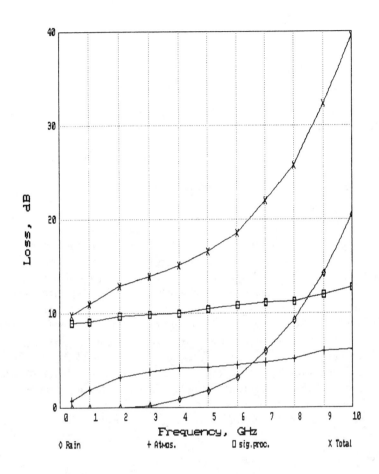

Figure 16.2: Example active aperture system losses for an airborne air-to-air application.

For a dry radome:

$$\alpha_R = 0.82 + 0.24(F_{GHz}/15)^2, \qquad dB \ one-way \qquad (16.8)$$

For a wet radome:

$$\alpha_R = [0.821 + 2.056\sin(\phi_e)]\,2.95\log(F_{GHz}/4.393) \times 10^{(1/3)\log(R/4)}, \qquad dB \ one-way \qquad (16.9)$$

where

ϕ_e is the elevation angle to the target R is the rain rate in mm/hr

The approximations appear valid for the frequency range of about 9 to 14 GHz.

Loss due to quantization noise added by A/D converter

From Hansen, [18], the quantization of the A/D sampling relative to the noise level, k, causes an increase in the total noise. This increase is a loss:

$$L_{AD} = 10 \log[1 + (12k^2)^{-1}], \text{ dB} \tag{16.10}$$

Range straddle loss as a function of sample rate

The average straddle loss as a function of the A/D sampling time interval relative to the half-power pulsewidth, τ_3, after Hansen [18] is:

$$L_s \approx -4.45 + 2.12(\Delta + 0.6) + 2.35/(\Delta + 0.6), \qquad \text{dB for } \Delta > 0.3 \tag{16.11}$$

where Δ is the A/D sampling time interval normalized by τ_3.

Collapsing losses in digital radar detection

Collapsing losses are the signal-to-noise ratio losses due to the addition of noise samples to one or more samples of signal-to-noise ratio. For analog processing systems the losses occur for the sum of the amplitudes of N_c channels followed by video integration. For digital detection systems DiVito and Galati, [17], predict the collapsing losses for schemes where the 1-bit outputs of the N_c channels are logically added by the OR operator. A coherent Doppler radar where there is only a signal in one filter and independent noise in each filter and the output of all filters are logically combined with an OR operator will have the collapsing loss as shown in Figure 16.3.

When a binary moving window is used to integrate the output of N_c channels the collapsing losses shown in Figure 16.4 and Figure 16.5 can be expected. Figure 16.4 is for a fixed target and Figure 16.5 is for a fast fluctuating target, Swerling 2. The data are shown for N cells of the binary moving window. The solid curves in each figure are for use of the moving window prior to the OR operation. The dashed curves are for use of the moving window after the OR operation.

MTI noise integration loss

MTI signal processing correlates the receiver noise which results in a loss. The loss is shown in Tables 16.4 and 16.5.

Table 16.6: MTI noise integration loss. After Trunk [16].

N-pulse MTI	2	3	4	5
Loss, dB	1.0	1.8	2.2	2.5

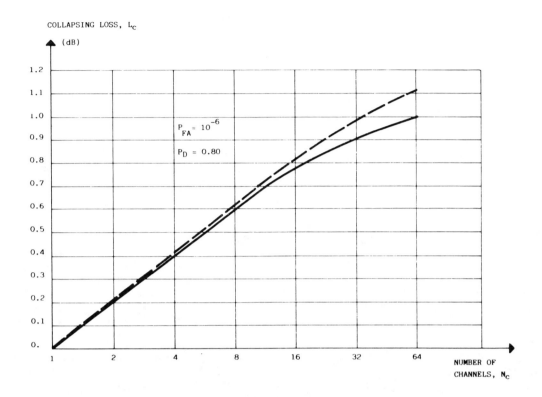

Figure 16.3: Collapsing loss for logically combining with an OR operator the output of N_c Doppler filters or the threshold output of N_c noncoherently integrated channels. $P_d = 0.8$ and $P_{fa} = 10^{-6}$. Solid line: fixed target. Dashed line: slowly fluctuating target, Swerling 1. From DiVito and Galati, [17].

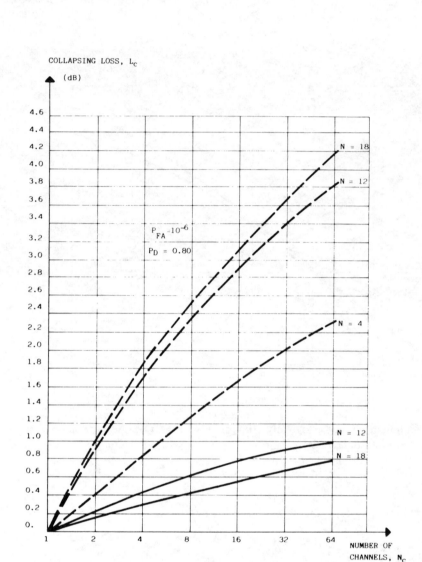

Figure 16.4: Collapsing loss for logically combining with an OR operator the output of N_c channels that use a binary moving window. Data for a steady target. $P_d = 0.8$ and $P_{fa} = 10^{-6}$. Solid lines; moving windows before OR operation. Dashed line; moving window after OR operation. From DiVito and Galati [17].

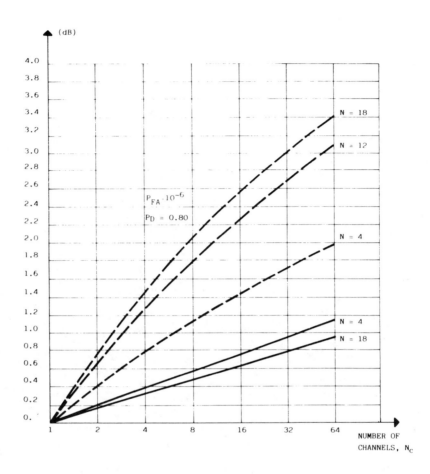

Figure 16.5: Collapsing loss for logically combining with an OR operator the output of N_c channels that use a binary moving window. Data for a fast fluctuating target, Swerling 2. $P_d = 0.8$ and $P_{fa} = 10^{-6}$. Solid lines; moving windows before OR operation. Dashed line; moving window after OR operation. From DiVito and Galati [17].

Bibliography

[1] Ausherman, D.A. (ed.) "Cumulative Index on Radar Systems for 1985–1989," IEEE Trans. AES-27, no. 3, May 1991, follows p.580.

[2] Blake, L.V., *Radar Range-Performance Analysis*, Artech House, Norwood, MA, 1986.

[3] Barton, D.K., Radars, Volume 2 - *The Radar Equation*, Artech House, Norwood MA, 1974.

[4] Barton, D.K., C.E. Cook, P.C. Hamilton, *Radar Evaluation Handbook*, Artech House, Norwood, MA, 1990.

[5] Skillman, W.A., *AIRCOVER: Airborne Radar Vertical Coverage Calculation Software and Users Manual*, Artech House, Norwood, MA, 1990.

[6] Barton, D.K., W.F. Barton, *Radar evaluation software*, Artech House, Norwood, MA, 1991.

[7] Barton, D.K., W.F. Barton, *Modern Radar System Analysis Software and User's Manual*, Artech House, Norwood, MA, 1990.

[8] Fielding, J.E., G.D. Reynolds, *RGCALC: Radar Range Detection Software and User's Manual*, Artech House, Norwood, MA., 1987.

[9] Skillman, W.A., *SIGCLUT: Surface and Volumetric Clutter-to-Noise, Jammer and Target Signal-to-Noise Radar Calculation Software and User's Manual*, Artech House, Norwood, MA., 1987.

[10] Fielding, J.E., G.D. Reynolds, *VCCALC: Vertical Coverage Plotting Software and User's Manual*, Artech House, Norwood, MA., 1988.

[11] CAE Soft Corp., *CAE Soft-Interactive Radar Simulation System*, Rockwall, TX , December 1989.

[12] Long, W.H., D.H. Mooney, W.A. Skillman, "Pulse Doppler Radar," Ch. 17 *Radar Handbook*, M. Skolnik, ed., McGraw-Hill, New York, 1990.

[13] Clarke, J. "Airborne Early Warning Radar," Proc. IEEE, February 1985.

[14] Barton, D.K., Radars, vol. 7, *CW and Doppler Radar*, Artech House, Norwood, MA, 1980.

[15] Kim, Y.S., W.A.Sandberg, "A methodology for computing link availability," IEEE Trans. AES-23, no. 2, pp. 255–262.

[16] Trunk, "MTI noise integration loss," IEEE Proc., vol. 65, no. 11, November 1977, pp. 1620-1621.

[17] DiVito, A., G. Galati, "Evalutation of collapsing losses in digital radar detection," IEEE Trans. AES-21, no. 2, March 1985, pp. 266–271.

[18] Hansen, V. Gregers, "Topics in Radar Signal Processing," Microwave Journal, March 1984, pp. 24–44.

[19] Morchin, W.E., *Airborne Early Warning Radar*, Artech House, Norwood, MA, 1990.

Notes

Notes

Chapter 17

Resolution

17.1 Introduction

Resolution is a measure of how far apart two signals must be before they can be distinguished from each other. The ability to distinguish one signal from another is a function of the shapes of the two signals. A common measure is the half-power width of the signal dimension, whether angle, time, range, or frequency, [2]. This common measure and other similar ones such as noise bandwidth, and ambiguity function half-power width are aperture limited, [1]. Aperture is used in the broad sense as it applies to either angle, time, range, or frequency.

There are signal processing techniques available to improve upon the aperture limited performance. The resolution so obtained is called *superresolution*. Typically, superresolution provides 3 to 8 times improvement in aperture limited resolution.

17.2 Aperture Limited Resolution

We continue with the broad definition of aperture as given in the introduction. Following the method shown by Carpentier, [3] for two Gaussian signals, without noise, added together; we require two maxima as a definition of resolution. The minimum separation of two such signals as a function of the amplitude, $A \leq 1$, of one signal relative to the other is shown in Table 17.1.

Figure 17.1 illustrates the combined amplitude of the two Gaussian signals for the relative amplitudes listed in Table 17.1. Each curve is plotted in descending order corresponding to the relative amplitudes listed. The values of x/X are those listed in Table 17.1 plus 0.1. We can see that noise on either or both signals will degrade resolution and that such noise should be at least 15 dB less than the weaker of the two signals to avoid such degradation.

17.3 Superresolution[1]

Superesolution describes the application of parametric spectrum estimation and statistical parameter estimation methods to the problem of resolution [7]. Resolution in angle and Doppler frequency are the most common applications. Prior knowledge about the signals to be resolved is used to

---

[1] This section was coauthored with David C. Lush.

Table 17.1: Difference between two Gaussian signals (angle, time, or frequency) required before such signals can be resolved as a function of the relative amplitude between the two signals. After Carpentier, [3].

Relative amplitude	$x > kX$ $k =$
1	1.1
0.5	1.4
0.1	1.9
0.01	2.3
0.001	2.6

X is the half-power width of a Gaussian signal.

develop an estimate of the unknown parameters of a signal. Superresolution techniques are not generally applicable directly to detection except for restrictive applications where detection is a problem of resolving target and interference signals. For example, a superresolution technique might be used to decrease the minimum velocity difference required for target detection between mainbeam clutter and target.

The use of superresolution depends on using a representative model of signals to be resolved and a sufficiently high signal-to-noise ratio, normally higher than that required for detection.

Two types of models are used for superresolution; rational transfer functions and sinusoidal signal functions. Rational transfer functions are used to resolve multiple targets in the presence of clutter with Doppler spread from one another. Sinusoidal signal functions are used for resolving angle responses of signals with multiple element antennas.

17.3.1   Rational Transfer Function Model

The rational transfer function model consists of a linear filter with signal and inteference, if it exist, as input. The filter may be:

1. all-pole: autoregressive – most appropiate when the signal consists of a small number of signal samples from those available from narrowband components in noise,

2. all-zero: moving average – most appropriate for broadband signal components,

3. pole-zero: autoregressive moving average – most appropriate when both narrowband and broadband signal components are expected.

Estimates of the signal autocorrelation formed from the signal samples are used to estimate the filter parameters. A spectrum estimate is found by evaluating an analytical exspression for the filter frequency response.

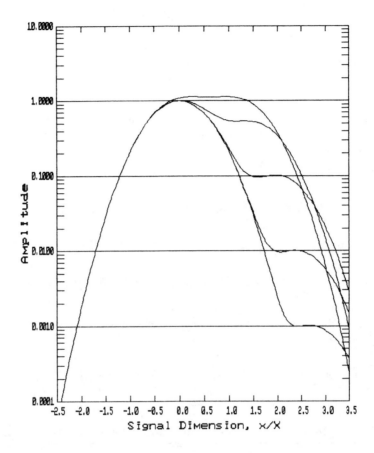

Figure 17.1: Combined amplitude of two Gaussian signals free of noise. See text for explanation of correspondence between the parameters used in Table 17.1 and the curves in the figure.

17.3.2 Sinusoidal Signal Model

The sinusoidal model consists of a number of sinusoids for which the frequencies, amplitudes and phases are determined by an eigenanalysis. The eigenanalysis is performed on the autocorrelation matrix of the signal samples. The elements of the autocorrelation matrix include signals and interference. In one such process, MUltsiple SIgnal Classification (MUSIC), [8], the signal frequencies are estimated as peaks of a scaler function of the signal associated eigenvectors. In the case of narrowband signals, signal-to-noise ratio can be increased prior to the superresolution process by executing an FFT on each array element signal.

17.3.3   Spatial Estimation Using Multiple Antenna Beams

Mayhan and Niro [9] have determined, when using maximum entropy spectral analysis, that the uncertainity of locating the angular position of a signal source in the antenna mainbeam is a function of the accuracy to which the antenna radiation patterns are known. Their graphical results that show the uncertainity radius in units of the antenna beam half-power beamwidth for locating a single source with a signal-to-noise of 40 dB can be fit with:

$$u_{BW} \approx 46\sigma^{1.43} \tag{17.1}$$

where σ is the rms error of the aperture illumination function.

Likewise the resolution between two equal 40 dB SNR sources as limited by the aperture illumination function errors can be fit with:

$$\delta_{BW} \approx 2.24\sigma^{0.47} \tag{17.2}$$

where δ_{BW} is in units of the half-power beamwidth.

17.4   Resolution Required

17.4.1   Target Interpretation, Probability of Recognition

Table 17.2 summarizes the resolution required for detection and interpretation of some typical radar targets. Table 17.3 predicts the probability of recognizing a target for various ratios of target and resolution dimensions.

Table 17.2: Resolution required for interpretation tasks, meters. After [10].

Object	Detection	Recognition	Precise identification	Description	Technical intelligence
Missile sites	3	0.5	0.6	0.3	0.08
Radar	3	0.9	0.3	0.15	0.04
Aircraft	4.5	1.5	0.9	0.15	0.03
Nuclear weapons Components	2.4	1.5	0.3	0.03	0.01
Surface submarines Command	30.5	6	1.5	0.9	0.03
Headquarters	3	1.5	0.9	0.15	0.03
Vehicles	1.5	0.6	0.3	0.05	0.03

Table 17.3: Probability of target recognition. After Mrstik [4].

Ratio of target dimension to resolution	Probability of recognition
18–24	1.0
cf 12–16	0.95
9–12	0.80
6–8	0.50
4.5–6	0.30
3–4	0.10
1.5–2	0.02

Bibliography

[1] Richards, M.A., "Iterative noncoherent angular superresolution," IEEE Nat. Radar Conf. Proc., Ann Arbor, MI, April 20–21, 1988, pp. 100–105.

[2] Barton, D.K., H.R. Ward, *Handbook of Radar Measurement*, Artech House, Norwood, MA, 1984.

[3] Carpentier, M.H., *Principles of Modern Radar Systems*, Artech House, Norwood, MA, 1988.

[4] Mrstik, A.V., et al., "RF systems in space," RADC-TR-83-91 Vol.2, AD-A133735, April 1983.

[5] Bullard, B.D., P.C. Dowdy, "Pulse Doppler signature of a rotary-wing aircraft," IEEE Nat. Radar Conf. RADAR-91, March 12–13, 1991, pp. 160–163.

[6] Chiavetta, R.J., "Target scattering fundamentals," Ch. 2 *Airborne Early Warning Radar*, ed. W.C. Morchin, Artech House, Norwood, MA, 1990.

[7] Kay, S.M., *Modern Spectral Estimation: Theory and Applications*, Prentice-Hall, Englewood Cliffs, NJ, 1988.

[8] Schmidt, R.O., "A Signal Subspace Approach to Multiple Emitter Location and Spectral Estimation," Ph.D. dissertation, Stanford University, 1981.

[9] Mayhan, J.T., L. Nuro, "Spatial spectral estimation using multiple beam antennas," IEEE Trans. AP-35, no.8, August 1987, pp. 897-906.

[10] IEEE Spectrum, July 1986.

Notes

Notes

Chapter 18

Search Radar

18.1 Search Equation for Cumulative Probability of Detection

In Chapter 16 we presented the usual form for the search radar equation. The basis for that equation was the signal-to-noise ratio for a single-scan probability of detection. Here, a form of the radar equation that is useful for determining the average power-aperture product for a specified cumulative probability of detection is given [1]:

$$P_{avg}A_e = \frac{4\pi k T F_n L V_t \Omega R^3}{(R/R_1)^3 Q_i^3(N)\sigma} \qquad (18.1)$$

$Q_i^3(N)$ $\quad$ =1 for coherent integration

$Q_i^3(N)$ $\quad$ $\approx 1 - 0.12 \log(N)$, $\quad 1 \le N \le 4$

$\qquad\qquad \approx 1.05 - 0.18 \log(N)$, $\quad 5 \le N \le 1000$, after [3]

N $\qquad$ is the number of pulses integrated

V_t $\qquad$ is the target radial velocity

Ω $\qquad$ is the angular search volume, sr

R $\qquad$ is the range for P_{cum}, the cumulative probability of detection

R_1 $\qquad$ =$(R_0^4/\Delta R)^{1/3}$

R_0 $\qquad$ is the range for which the integrated $(S/N) = 0$

ΔR $\qquad$ is the radial range change toward the radar between scans

The above value of $Q_i(N)$ is for a Swerling 1 target and for $n = 10^6$, Marcum's false number, the number of noise samples in a time in which the probability is 1/2 that the noise will exceed

367

the threshold. See [2] for more cases of n and also for a Swerling 3 target. The minimum radar resources required will occur when R/R_1 is maximum, after [2] with approximation by [3]:

$$(R/R_1) \equiv (R/R_1)_{\text{max}} \approx 0.1 + 0.1 \ln[1/(1 - P_{\text{cum}})]^A , 0.5 \leq P_{\text{cum}} \leq 0.99 \qquad (18.2)$$

$A = -0.84$ for Swerling 1 target
$\quad = -0.8$ for Swerling 3 target
$\quad$ Corresponding ΔR normalized by R is:

$$(\Delta R/R)_{\text{max}} \approx B \ln[1/(1 - P_{\text{cum}})]^A]$$

$B = 0.37$ for Swerling 1 target
$\quad = 0.43$ for Swerling 3 target

The time to search is:

$$T_s = (\Delta R/R)_{\text{max}}(R/V_t) \qquad (18.3)$$

If T_s is found to be too long using $(\Delta R/R)_{\text{max}}$, choose a lower value for $(\Delta R/R)$ and solve for a new (R/R_1) using

$$\frac{(\Delta R/R)}{(\Delta R/R)_{\text{max}}} \approx C \left[\frac{(R/R_1)}{(R/R_1)_{\text{max}}} \right]^D$$
$$0.35 \leq \frac{(R/R_1)}{(R/R_1)_{\text{max}}} \leq 0.7 \qquad (18.4)$$

$C = 0.26$ for a Swerling 1 target
$\quad = 0.33$ for a Swerling 3 target

$D = 4.75$ for a Swerling 1 target
$\quad = 4.1$ for a Swerling 3 target

18.2   Apportionment of Power and Time

18.2.1   Programmed Time and Power

For an agile antenna beam such as in a phased array radar system it is possible to vary the program of power and time use. For minimum power use, [3] section 8.4, the time spent for each function is:

$$t_1 = T/(k_1 + k_2\sqrt{k_1 E_2/k_2 E_1} + \ldots + k_i\sqrt{k_1 E_i/k_i E_1}$$
$$t_2 = t_1\sqrt{k_1 E_2/k_2 E_1}$$
$$\vdots$$
$$t_i = \sqrt{k_1 E_i/k_i E_1} \qquad (18.5)$$

where T is the total time for performing all the functions, E_i is the radar energy for the ith radar function, and $k_i = T/T_i$. T_i is the time spent on the ith function. Usually at least one function is performed only once during T, let that be function 1; so that $k_1 = 1$. Other $k_i = nk_1$, where n is the number of more times the ith function is repeated than the first.

Time, rather than power, can be minimized by interchanging P_i with t_i where the total power, P, is a constraint.

18.2.2   Search Time Savings with Sequential Detection

The time spent searching with a dual-threshold system as compared to a single-threshold system is, after Hammerle, [3]:

$$\bar{T}_{SD}/T_S = C_0 + N_{cd}P_{fD}^{r_o} \qquad (18.6)$$

$\bar{T}_{SD}$ is the average time spent in search whith a dual-threshold system with targets present T_S is the time spent in search with a single threshold system.

$C_0 \approx 1 - \exp\left\{-10^{[0.25+0.29B+(0.054-0.036B)\log N_{cd}]}\right\}$

B is ratio of time when integration is not performed to time of target observation.

N_{cd} is the number of range-velocity cells in a beam spot.

$r_o \approx 0.43 - 0.024n + (0.256 - 0.015n)\log N_{cd}$

n is given by $P_{fD} = 10^{-n}$

The probability of false alarm in a dual-threshold system is the product of the false alarm probabilities for the 1 and 2 thresholds.

$$P_{fD} = P_{f1}P_{f2}$$

Note that the time to search with a dual-threshold system may exceed that for a single threshold system for large number of range-velocity cells, certain values of probability of crossing the first threshold, and dead time.

18.2.3   Radar Time Use

Radar time use is the ratio of beam spot dwell time to the revisit time to a given spot in angular space multiplied by the number of beam spots:

$$t_u = T_{ob}N_d/T_r \qquad (18.7)$$

where T_{ob} is the dwell time at a beam spot, N_d is the number of beam spots, and T_r is the revisit time. It is possible that $t_u \geq 1$ providing there is more than one antenna beam and more than one receiver-processor.

18.3   Search Time, Antenna Area, and Target Velocity Trades

18.3.1   Target Velocity Resolution

The velocity resolution of a pulse Doppler radar is [3]:

$$\Delta v = \frac{\Omega A_e}{2T_s\lambda} \qquad (18.8)$$

where Ω is the angular search volume, A_e is the effective antenna aperture, and T_s is the time to search Ω. We see that reducing A_e improves velocity resolution, but that poorer detection performance results because of reduced power-aperture product.

18.3.2   Minimum Detectable Target Velocity

The minimum detectable target velocity of a pulse Doppler radar when mainbeam clutter has been attenuated to be equal to the target signal is [3]:

$$V_{\min} = \frac{\sigma_c}{0.84} + \frac{\Omega A_e}{4T_s \lambda}$$

σ_c is the rms main beam clutter velocity spread

σ_c^2 $= \sigma_p^2 + \sigma_{cl}^2 + \sigma_f^2$

σ_p standard deviation of clutter spread due to platform velocity, if any;

σ_{cl} is standard deviation of clutter internal motion

σ_f is standard deviation of the mainbeam clutter filter

Bibliography

[1] Rusnak, I., I.Gertner "Optimization of surveillance radar with instrumented range," IEEE Trans. AES-23, no.5, September 1987, pp. 712–715.

[2] Mallett, J.D., L.E. Brennan, "Cumulative probability of detection for targets approaching a uniformly scanning search radar," IEEE Proc., vol. 51, no. 4, April 1963, pp. 596–601, with corrections IEEE Proc., vol. 52, no. 6, June 1964, pp. 708–709.

[3] Morchin, W.C., *Airborne Early Warning Radar*, Artech House, Norwood, MA, 1990, section 8.3.

Notes

Notes

Chapter 19

Signal Design

19.1 General

19.1.1 Time-Frequency Hop

Bellegarda and Titlebaum, [1], present a procedure for determining the sequence of frequencies to be used in a frequency hopped system to achieve both good auto- and cross-ambiguity properties. Their procedure is:

$$f_k = f_0 + y_k B/N, \qquad k = 0, 1, \ldots, N-1$$

$$y_k = modN \text{ of } \left[a\frac{k(k+1)}{2} \right], \qquad 0 \le k \le \frac{N-1}{2}$$

$$y_k = modN \text{ of } \left[b\frac{k(k+1)}{2} + (a-b)\frac{N^2-1}{8} \right], \qquad \frac{N-1}{2}, \le k \le N-1$$

(19.1)

where N is the number of hops in a sequence, B is the bandwidth of the signal, and a and b are each any number of the set $1, 2, \ldots, N-1$.

19.1.2 Resolution of a Target Dimension

To resolve by autocorrelation an incremental distance, δx, on a traveling object one can use a frequency, F, or a frequency difference, ΔF [2]:

$$F = c/2\delta x$$
$$\Delta F = c/2\delta x$$

(19.2)

where c is the propagation velocity.

19.2   Frequency Modulated

19.2.1   Component Phase Errors

The relative sidelobe level due to nonlinearities of the phase response of components such as the sideband modulator or frequency multiplier of a FM radar will cause narrowband phase modulation a pair of sidebands of relative sidelobe level of [4]:

$$SLL = \frac{\tau \pi}{\Delta f} \frac{df}{dt} \Phi_p \qquad (19.3)$$

where τ is the time delay between the target echo and the receiver local oscillator. Δf is the frequency difference between phase departures from linearity, df/dt is the FM sweep rate, and Φ_p is the peak phase error.

19.2.2   Digital Generation of Linear FM Waveforms

Random and discrete sidelobe levels

Postema [16] analyzed the effect of deterministic and random errors on the attainable range sidelobe level in a coherent linear FM radar waveform, as used in high resolution radar. The form of the radar is shown in Figure 19.1. The deterministic range sidelobe levels attainable for various multiplier factors and time-bandwidth products is shown in Figure 19.2. The rms sidelobe levels attainable are shown in Figure 19.3 for various multiplier factors and number of bits in the D/A converter.

Griffiths and Bradford [4] give the following relationship for the relative sidelobe level of the first pair of phase quantization sidebands:

$$\begin{aligned} SLL &= 20 \log \frac{J_1(\beta)}{J_0(\beta)} \\ \beta &= 2M \tan^{-1}(2^b - 1)^{-1} \end{aligned} \qquad (19.4)$$

where $J_n(x)$ are Bessel functions of the first kind of order n and argument x.

19.2.3   Frequency-hopped FM Segments

Frequency-hopped FM segments of a wideband FM waveform can be used to avoid time-delay antenna beam steering. The antenna beam would be steered to a correct position for the next FM segment after completing the present segment. With non-periodic segments the range sidelobes will be [5]:

$$SLL \approx -(24 + \log N), dB \qquad (19.5)$$

where N is the number of segments in the wideband FM waveform.

19.2.4   Doppler Estimation Accuracy with Linear FM

The normal expression for Doppler estimation accuracy with a single linear FM:

$$\begin{aligned} \sigma_{f_d} &= \sqrt{\frac{B}{4\tau S/N}} \\ B &\quad \text{is the chirp bandwidth,} \\ \tau &\quad \text{is uncompressed pulsewidth,} \\ S/N &\quad \text{is the signal-to-noise ratio at the peak of the compressed pulse.} \end{aligned} \qquad (19.6)$$

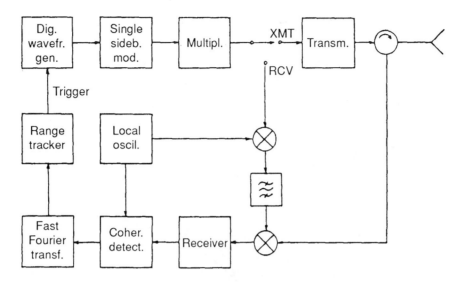

Figure 19.1: Conceptual block diagram of radar. From Postema [16].

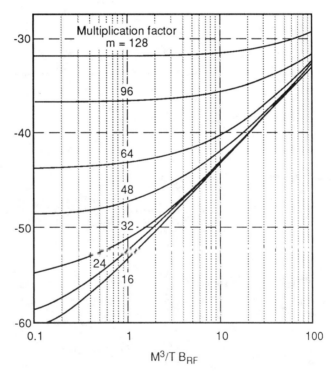

Figure 19.2: Discrete sidelobe level for a digitally generated linear FM waveform. From Postema [16].

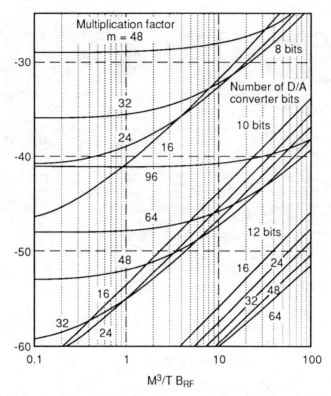

Figure 19.3: RMS sidelobe level for a digitally generated linear FM waveform. From Postema [16].

was found by Daum, [6], to be valid for high S/N, such as:

$$S/N \geq 3\pi^2(B\tau)^2/8 \tag{19.7}$$

A more general relationship is:

$$\begin{aligned}
\sigma_{f_d} &= \frac{B}{[k(S/N)^{3/2}\tan^{-1}(n\sqrt{a})]^{1/2}} \\
k &= 8\sqrt{2}/(\pi\sqrt{3}) \\
n\sqrt{a} &= B\tau\pi\sqrt{3/8}/\sqrt{S/N}
\end{aligned} \tag{19.8}$$

To account for the multiple target environment Daum suggests a way to estimate a target Doppler of interest by using $\pm M$ phase samples nearest the peak of the compressed pulse. M is $\ll B\tau/2$. For this design the term $n\sqrt{a}$ is replaced by:

$$M\sqrt{3\pi^2/(2S/N)}$$

19.3   Phase Coded

19.3.1   Biphase Codes

Marvin and Philip Cohen, [8], reference a search procedure for determining filter weights of a mismatched filter to Barker and other biphase codes for achieving minimum integrated sidelobe levels and low loss of processing gain. Marvin Cohen, [7] presented these weights for the Barker codes, shown in Table 19.1. The Cohen and Cohen data also show that the integrated and peak sidelobe levels of the 13-bit Barker code decreased about 2/3 dB for each added filter cell for filters of length 23 to 63. The peak sidelobe level averaged about 8 dB lower than the integrated level.

Peak sidelobe levels for the Cohen and Cohen minimum peak sidelobe level codes are shown in Figure 19.4. Also shown are their results for various mismatched filter lengths to the 13-bit Barker code.

19.3.2   Quadraphase Codes from Biphase Codes

Tayler and Blinchikoff, [9], desribe a method for creating a quadraphase code from a biphase one. The quadraphase code used 0, 90, 180, and 270 phase values that were used in the place of biphase values of 0 and 180. Their conversion followed:

$$V_k = j^{s(k-1)} W_k \, , k = 1, 2, \ldots N \tag{19.9}$$

where s is fixed at either $+1$ or -1, N is the number of subpulses, and W_k is the phase of the biphase code. They converted five biphase codes of lengths; 13, 25, 27 and (2) of 28. Their codes are shown in Figure 19.5. Their code used subpulses of half-cosine shape which results in 12 dB/octave spectrum falloff rather than rectangular shape which results in 6 dB/octave. They also point out, [10], that for acceptable performance, applications of biphase and quadriphase codes are limited to uncompressed pulse lengths less than $0.3/f_{D_{max}}$, where $f_{D_{max}}$ is the maximum Doppler shift of a moving target.

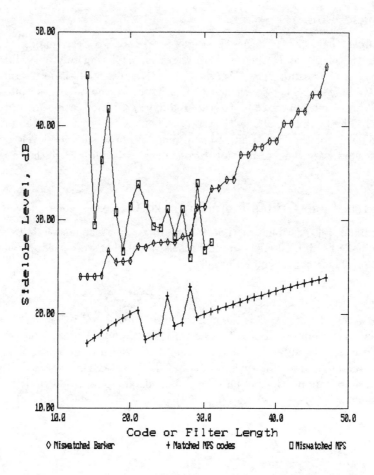

Figure 19.4: Peak sidelobe level of mismatched filtering of 13-bit Barker code and matched and mismatched filtering of Cohen and Cohen minimum peak sidelobe level biphase codes, after [7].

Table 19.1: Mismatched filter weights for integrated sidelobe levels
for the Barker codes, after [7].

	Barker Code Length					
	3	4	5	7	11	13
PSL, dB	−37.6	−30.1	−35.5	−29.9	−27.2	−33.4
ISL, dB	−32.5	−28.2	−28.9	−21.2	−17.1	−25.4
LPG, dB	1.26	1.65	0.61	1.25	0.71	0.20
Weight number	**Weight**					
a_1	0.029	0.082	0.064	0.073	0.126	−0.115
a_2	0.048	0.129	0.036	0.084	−0.042	−0.124
a_3	0.088	0.043	−0.082	0.101	0.188	−0.048
a_4	0.143	−0.198	−0.214	0.107	−0.125	−0.04
a_5	0.236	−0.391	−0.236	0.143	0.131	−0.115
a_6	0.381	−0.239	0.036	0.198	−0.208	−0.093
a_7	0.618	0.36	0.581	0.268	0.188	−0.045
a_8	1	1	*	0.282	−0.291	−0.146
a_9	−0.618	0.882	*	0.223	0.126	0.025
a_{10}	0.381	−0.48	*	0.187	−0.256	0.618
a_{11}	−0.236	0.262	*	0.335	0.758	0.842
a_{12}	0.143	−0.141	−0.036	0.724	0.417	0.68
a_{13}	−0.088	0.077	−0.236	1	0.737	0.839
a_{14}	0.048	−0.042	0.214	0.676	−0.993	0.638
a_{15}	−0.029	0.023	−0.082	−0.287	−0.684	−0.812
a_{16}		−0.012	−0.036	−0.676	−1	−1
a_{17}		0.007	0.064	1	0.684	0.812
a_{18}		−0.004		−0.724	−0.993	0.638
a_{19}				0.334	−0.737	−0.839
a_{20}				−0.187	0.417	0.68
a_{21}				0.223	−0.758	−0.842
a_{22}				−0.282	−0.256	0.618
a_{23}				0.268	−0.126	−0.025
a_{24}				−0.198	−0.291	−0.146
a_{25}				0.143	−0.188	0.045
a_{26}				−0.107	−0.208	−0.093
a_{27}				0.101	−0.131	0.115
a_{28}				−0.084	−0.125	−0.042
a_{29}				0.073	−0.188	0.048
a_{30}					−0.042	−0.124
a_{31}					−0.126	0.115

PSL is the peak sidelobe level relative to the peak of the mainlobe response

ISL is the ratio of the power in the sidelobes to the signal power

LPG is the loss of processing gain

* values not given in reference

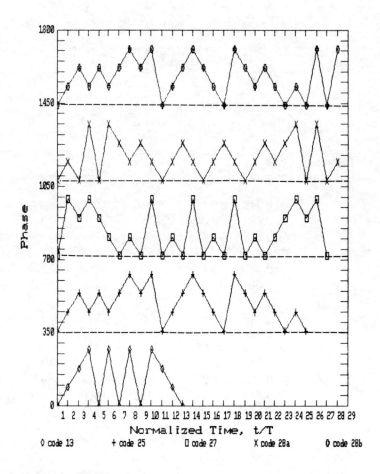

Figure 19.5: Quadraphase codes as derived from biphase codes. After Taylor and Blinchikoff, [9]. The phase scale has been offset 360° between codes to separate the graphs. Each code starts at 0°, t is the subpulse lenth, and $T = Nt$, where N is the number of subpulses.

The salient features of their codes are shown in Table 19.2.

Table 19.2: Characteristics of codes, after [9].

Code Peak/sidelobe ratio, dB	22.2	21.6	19	22.4	22.4
Energy pulse width, multiple of t	1.22	1.32	1.30	1.34	1.34
Range sampling loss, dB	0.78	0.79	0.77	0.74	0.76
Compression ratio for 6 dB pulse width	8.6	16.3	17.2	17.7	18

The ambiguity function of two quadraphase codes is shown in Figures 19.6 and 19.7.

19.3.3   Polyphase Codes

The P3 and P4 codes as presented by Lewis and Kretschmer [12] [13] used for creating quadratic phase modulation can be described by:

For the P3 code:

$$S_i = \exp(j\pi i^2/N), 0 \leq i \leq N-1 \tag{19.10}$$

For the P4 code:

$$S_i = \exp(j\pi i^2/N + j\pi i), 0 \leq i \leq N-1 \tag{19.11}$$

where the phase is module 2π.

Wang and Shyu, [14], give a technique for implementing the P3 and P4 codes that is based on matrix extension of the Frank code.

Luke, [15] presents sequences with perfect periodic correlation for sequence lengths of 3 to 60. A perfect periodic sequence is one in which the autocorrelation function gives a peak response equal to the sum of the squares of the values in the sequence for $n = 0$ in the region $0 \leq n \leq N$, where N is the total number of values in a sequence and gives a value of 0 for $n \neq 0$. The first 10 Luke sequences are shown in Table 19.3. The reader is referred to [15] for the remaining sequences to $N = 60$. η is the energy percentage which is the sum of the squares of the sequence values divided the product of N and the square of the maximum value in a sequence.

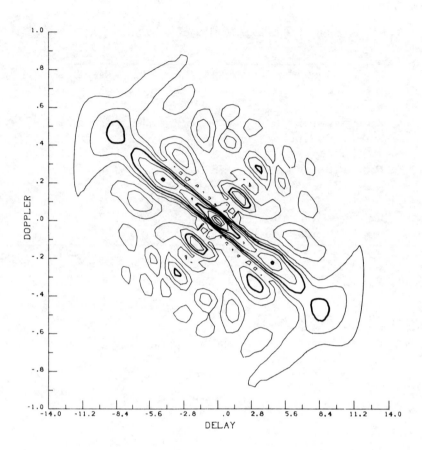

Figure 19.6: Ambiguity function for 13-bit quadriphase pulse, contours begin at 0.1 and are shown for increments of 0.1. Contours 0.3 and 0.6 are emphasized. Doppler axis is from $-1/T$ to $+1/T$ and delay axis is from $-14T$ to $+14T$. From Levanon and Freedman [11].

Table 19.3: The first 10 Lüke perfect sequences.

N	η	Sequence
3	75	1, 1, −0.5
4	100	1, 1, 1, −1
5	67	1, −1, 0.33, 1, 0.5
6	67	1, 0, −1, 1, 0, 1
7	72	1, 1, 1, −0.586, 1, −0.586, −0.586
8	69	1, 1, 0.17, 1, −0.17, 1, −1, −0.66
9	71	1, −1, −0.45, 1, 1, 0.68, −0.85, 0.12, −0.98
10	66	1, −1, −0.87, −0.24, −0.87, 0, −0.87, 0.24, −0.87, 1
11	77	1, −1, 1, 0.9, −0.12, −1, 0.46, −1, 0.7, 1, 1
11	73	1, 1, −0.634, 1, 1, 1, −0.634, −0.634, −0.634, 1, −0.634

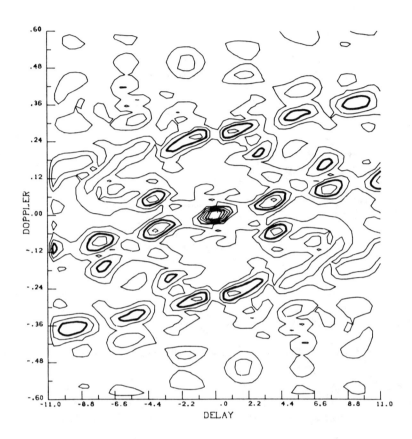

Figure 19.7: Ambiguity function for 28a quadriphase coded pulse, contours begin at 0.1 and are shown for increments of 0.1. Contours 0.3 and 0.6 are emphasized. Doppler axis is from $-1/T$ to $+1/T$ and delay axis is from $-14T$ to $+14T$. From Levanon and Freedman [11].

19.4 Pulse Burst

19.4.1 Range Determination in High-PRF Pulse-Doppler Radar

The reader is referred to Postema [16] for a table construction procedure to reduce the computations required to determine range in a high PRF pulse-doppler radar.

19.4.2 Multiple PRF Ghosting and Probability of Target Visibility in Ambiguous Range Systems

Nevin, [17], presents a procedure for determining an estimate of the ghosting probability for a multiple-PRF pulse doppler radar. His procedure is:

1. Determine the visibility of an arbitrary target in one PRF as:

$$V_1 = \left(1 - \frac{2TB}{\overline{N_c}}\right)\left(1 - \frac{2f_{Da}}{\overline{PRF}}\right) \qquad (19.12)$$

where

TB is the waveform time-bandwidth product, or expanded pulse width in range cells

$\overline{N_c}$ is the mean number of range cells per PRF

f_{Da} is the minimum visible ambiguous Doppler frequency

$\overline{PRF}$ is the mean PRF

2. Determine the probability that an arbitrary target is visible in exactly M of N PRFs as:

$$V_{M \, of \, N} \approx \binom{N}{M} V_1^M (1 - V_1)^{N-M} \qquad (19.13)$$

Bibliography

[1] Bellegarda, J.R., E.L. Titlebaum, "Time-Frequency Hop Codes based upon extended quadratic congruences," IEEE Trans. AES-24, no. 6, November 1988, pp. 726–742.

[2] Gjessing, D.T., "Matched illumination target adaptive radar for challenging applications," IEE Int. Radar Conf., London, October 19-21, 1987, pp. 287–291.

[3] Postema, G.B., "Generation and performance analysis of wideband radar waveforms," IEE Int. Radar Conference, London, October 19-21, 1987, pp. 310–314.

[4] Griffiths, H.D., W.J. Bradford, "Digital generation of wideband FM waveforms for radar altimeters," IEE Int. Radar Conf., London, October 19-21, 1987, pp. 325–329.

[5] Maron, D.E., "Non-periodic frequency-jumped burst waveforms," IEE Int. Radar Conf. Proc. 1987, pp. 484–488.

[6] Daum, F.E., "Doppler estimation accuracy of linear FM waveforms," IEEE Int. Radar Conf., Arlington, VA, 1985, pp. 107–112.

[7] Cohen, M.N., "Pulse compression in radar systems," Session B-3, IEEE Int. Radar Conf., Arlington, VA, May 8, 1990.

[8] Cohen, M.N., P.E. Cohen, "Biphase codes with minimum peak sidelobes," IEEE Nat. Radar Conf., Dallas, TX, March 29-30, 1989, pp. 62–66.

[9] Taylor, J.W., H.J. Blinchikoff, "The quadriphase code - A radar pulse compression signal with unique characteristics," IEE Int. Radar Conf., London, October 19-21, 1987, pp. 315–319.

[10] Taylor Jr., J.W., H.J. Blinchikoff, "Quadriphase code - A radar pulse compression signal with unique characteristics," IEEE Trans. AES-24, no. 2, March 1900, pp. 150–170.

[11] Levanon, N., A. Freedman, "Ambiguity function of quadriphase coded radar pulse," IEEE Trans. AES-25, no. 6, November 1989, pp. 848–853.

[12] Lewis, B.L., F.F. Kretschmer, "Linear frequency modulation derived polyphase pulse compression codes," IEEE Trans. AES-18, no. 5, September 1982.

[13] Kretschmer, F.F., B.L. Lewis, "Doppler properties of polyphase coded pulse compression waveforms," IEEE Trans. AES-19, no. 4, July 1983.

[14] Wang, C-C., H-C. Shyu, "An extended Frank code and new technique for implementing P3 and P4 codes," IEEE Trans. AES-25, no. 4, July 1989, pp. 442–448.

[15] Luke, H.D., "Sequences and arrays with perfect periodic correlation," IEEE Trans. AES-24, no. 3, May 1988, pp. 287–294.

[16] Postema, G.B., "Range ambiguity resolution for high PRF pulse-doppler radar," IEEE Int. Radar Conf., Arlington, VA, 1985, pp. 113–118.

[17] Nevin, R.L., "Waveform trade-offs for medium PRF air-to-air radar," IEEE National Radar Conference, Ann Arbor, MI, 1988, pp. 140–145.

Notes

Notes

Chapter 20

Signal Processing

20.1 Introduction

Signal processing is performed for the purpose of accomplishing certain functions. These functions are listed in Table 20.1. Berenz and Dunbridge, [1], suggest three divisions of signal processing: generation, extraction, and transformation. Signal generation includes modulation, signal systhesis, multiplication, amplification and beamforming. Signal extraction includes demodulation, demultiplexing, dehopping, filtering, detection, and imaging. Signal transformation includes frequency conversion, A/D conversion, amplification and delay.

We can identify four conceptual dimensions relating to signal processing. These four dimensions can be expressed with three levels of association. Starting with the highest level associations: we can consider the relationship between the functions that signal processsing must accomplish and the technology applied to achieve those functions. The second level is the association of implementation means of achieving the various processing techniques available. The third and final level is the association of devices used to accomplish an implemenatation. Table 20.1 illustrates a list of signal processing functions and techniques. The two can be interrelated by means of a matrix which would show that a number of techniques are combined to accomplish a certain signal processing function. The table similarly shows that various processing techniques can be accomplished by various methods. Finally, although not shown, a multitude of devices is available to construct an implementation.

This chapter is organized around the processing techniques. However, the subject of signal processing seems to have a nature to it that has prevented short distillations of the subject matter. Therefore we present an organized and annotated list of references in the text. This material is in addition to the references used for the technical material presented. We first present some bibliographical references for the target classification and recognition and imaging functions. This is followed by implementation and device references.

20.1.1 Target Classification/Recognition

1. Rothwell, E.J., K-M Chen, D.P. Nyquist, W. Sun, "Frequency domain E-pulse systhesis and target discrimination," IEEE Trans. AP-35, no. 4, April 1987, pp. 426–434.

   One of many references by the authors on target identification by means of identifying target resonance responses to designed pulse shapes.

Table 20.1: Listed functions of signal processing and techniques used to accomplish the functions, and methods of implementation.

Processing function	Processing technique	Implementation method
Signal enhancement	Sampling	Digital
Clutter suppression	Filtering	Analog
Interference suppression	- MTI	Acoustic
Target detection	- FFT/DFT	Optical
Target classification estimation	- Spectral	Acousto-optical
Imaging	- Correlation	
	- Pulse compression	
	Adaptive processing	
	- Antenna	
	- Doppler	
	- Neurological	
	Programmed processing	
	- General purpose	
	- Special purpose	
	- Hardwired	
	- Knowledge based	

2. Hua, Y., T.K. Sarkar, "A discussion of E-pulse method and Prony's method for radar target resonance retrieval from scattered field," IEEE Trans. AP-37, no. 7, July 1989, pp. 944–946.

Another version of the E-pulse method used to find target frequency response poles and subsequent identification of the target.

20.1.2   Imaging

1. Brooks, S.R., et al. "Real time SAR processing," Military Microwaves, London, England, 1986, pp. 411–416.

Equations for predicting the data flow rate into the range compression, azimuth compression and resampling and detection functions of a synthetic aperture radar are given. A spaceborne application example is used.

2. Tsao, J., B.D. Steinberg, "Reduction of sidelobe and speckle artifacts in microwave imaging: the CLEAN technique," IEEE Trans. AP-36, no. 4, April 1988, pp. 543–556.

A description of a technique used in radio astronomy which has been applied to obtaining good images of conventional radar targets with thinned arrays.

3. Chen, P., M. Haerle, "Monopulse azimuth resolution improvement applied to radar images," Military Microwaves, London, England, 1984, pp. 334–338.

Two methods of obtaining enhanced images by means of the use monopulse signals.

4. Li, H-J, N.H. Farhat, Y. Shen, "A new iterative algorithm for extrapolation of data available in multiple restricted regions with application to radar imaging," IEEE Trans. AP-35, no. 5, May 1987, pp. 581–588.

A method of modifying linear prediction of target responses across missing frequency data is described. The method is used to construct target images.

20.1.3   Digital Implementation of Signal Processing

An annotated bibliography of recent articles and papers pertinent to digital implementation of signal processing follows:

1. Aliphas, A., J.A. Feldman, "The versatility of digital signal processing chips," IEEE Spectrum, June, 1987, pp.40–50.

A tutorial on digital signal processing and a sampling of characteristics of available integrated circuits.

2. Rabinowitz, S.J., et al. "Applications of digital technology to radar," IEEE Proc. vol. 73, no. 2, February 1985, pp. 325–339.

A tutorial on the ways in which digital technology is used in radar to overcome the limits of analog circuits.

3. J. Roberts, P. Simpson, B. Merrifield, "Applying digital VLSI technoloby to radar signal processing," Microwave Journal, January 1986, pp. 129–142.

Digital signal processessing implemented with Very Large Scale Integration (VLSI) circuits is explained. The use of various solid state technologies are related to signal processing functions that can be implemented on a single chip.

4. Fair, R.B., "Challenges to manufacturing submicron ultra-large scale integrated circuits," IEEE Proc., vol. 78, no. 11, November 1990, pp. 1687–1705.

A look to the future with tabular and curve data giving current and future device density, temperature, defect and manufacturing tolerances.

20.1.4   Acoustic Device Implementation for Signal Processing

1. Kansy, R.J., et al. "Acoustic charge transport signal processors," Microwave Journal, November, 1988, pp. 141–152.

This reference describes device capabilities for supporting the various signal processing techniques. The analog implementations with the devices is said to have the flexibility of digital devices.

2. Ballato, A., T. Lukaszek, "Microwave acoustic material properties," Microwave Journal, May 1990, pp. 105–114.

Tabular data are given on frequency-temperature coefficients and material constants.

20.2   Sampling

20.2.1   Analog to Digital Converters

An approximate relation between the maximum number of bits b, and the A/D conversion rate, MHz, is, [9] Figure 7.6, [2] for 1990 technology:

$$b \approx Int(14.3 - 2.65 \log f_{\mathrm{MH}_z}) \tag{20.1}$$

where f_{MH_z} is A/D conversion rate in MHz.

20.2.2   Method for Reducing the Bandwidth of a Sampled Signal

Mitchell, [4], describes a procedure for prefiltering that efficiently reduces the bandwidth of a sampled signal. His procedure is to sample, or i.e. use, every other sample of a sequence of a values to be input to a low-pass FIR filter. The filtering operation would be:

$$\acute{x}_k = x_k + a_1(x_{k+1} + x_{k-1}) + a_2(x_{k+3} + x_{k-3}) + \ldots + a_n(a_{k+2n-1} + a_{k-2n+1}) \tag{20.2}$$

where n is the number of FIR filters multiplies, and k is the sample number index. Processing starts after $2n + 1$ samples are obtained. The performance of a filter length of $4n - 1$ is obtained with only n multiplies. He states that multiple stages of filtering by his method will achieve the same results with less multiplies as a single stage filter using full sampling.

20.2.3   Data Windowing

Often, it is desirable to apply amplitude weighting to data acquired to reduce the sidelobes of any transformations performed on such data. The weighting is applied to reduce abrupt changes at the beginning and end of acquired data. Table 20.2 shows some common time-data windows and their frequency-domain parameters. The major lobe height shown in the fourth column refers to the rectangular window as a reference. The last column gives the theoretical rate of decay of the sidelobes. The analogies between antenna, time-limited voltage waveforms, and frequency-limited amplitude spectral data as given by, [6] Table B.2, can be applied to the data in Table 20.2.

Table 20.2: Some common time-data windows and their frequency-domain parameters. β is the reciprocal of the windows time duration. From [5].

Unity Amplitude Window	Shape Equation	Frequency Domain Magnitude	Major Lobe Height	Highest Side Lobe (dB)	Band-width (3 dB)	Theoretical Roll-Off (dB/Octave)
Rectangle	$A=1$ for $t=0$ to T		T	-13.2	0.86β	6
Extended Cosine Bell	$A=0.5(1-\cos 2\pi 5t/T)$ for $t=0$ to $T/10$ and $t=9T/10$ to T; $A=1$ for $t=T/10$ to $9T/10$		$0.9\,T$	-13.5	0.95β	18 (beyond 5β)
Half Cycle Sine	$A=\sin 2\pi 0.5t/T$ for $t=0$ to T		$0.64\,T$	-22.4	1.15β	12
Triangle	$A=2t/T$ for $t=0$ to $T/2$; $A=-2t/T+2$ for $t=T/2$ to T		$0.5\,T$	-26.7	1.27β	12
Cosine (Hanning)	$A=0.5(1-\cos 2\pi t/T)$ for $t=0$ to T		$0.5\,T$	-31.6	1.39β	18
Half Cycle Sine³	$A=\sin^3 2\pi 0.5t/T$ for $t=0$ to T		$0.42\,T$	-39.5	1.61β	24
Hamming	$A=0.08+0.46(1-\cos 2\pi t/T)$ for $t=0$ to T		$0.54\,T$	-41.9	1.26β	6 (Beyond 5β)
Cosine²	$A=(0.5(1-\cos 2\pi t/T))^2$ for $t=0$ to T		$0.36\,T$	-46.9	1.79β	30
Parzen	$A=1-6(2t/T-1)^2+6|2t/T-1|^3$ for $t=T/4$ to $3T/4$; $A=2(1-|2t/T-1|)^3$ for $t=0$ to $T/4$ and $t=3T/4$ to T		$0.37\,T$	-53.2	1.81β	24

As additional information to the data in Table 20.2 Agoston and Henricksen give further data on the windows, [7]. They refer to a scallop loss, which is a measure of the flatness of the pass band response of the data window. They define the value of the scallop loss is the maximum value of deviation from peak window response to the response at the border with an adjacent window in an FFT application. They further define a worst case loss for windowing. This worst case loss is the addition of the scallop loss and the loss associated with sample weighting. Their values are shown in Table 20.3.

Table 20.3: Scallop and worst case loss for data windows.

Window	Scallop loss (dB)	Worst Case loss (dB)
Rectangle	3.92	3.92
Triangle	1.82	3.07
Hanning	1.42	3.18
Hamming	1.78	3.1
Blackman	1.1	3.47
Blackman-Harris	0.83	3.85

20.2.4 Quadrature Sampling

Relative level of sidebands

Errors in sampling or processing signals in quadrature networks cause sidebands [8]:

$$SB = -25 + 20 \log A_\epsilon$$
$$SB = -41 + 20 \log \Phi_\epsilon$$

(20.3)

where A_ϵ is the amplitude error, and Φ_ϵ is the phase error in degrees.

Annotated Bibliography

1. Chu, D. "Phase digitizing sharpens timing measurements," IEEE Spectrum, July 1988, pp. 28–32.

   A change of phase digitizer is used to extend the frequency range of phase coded chirp signals.

2. Liu, H., A. Ghafoor, P.H. Stockmann, "Time jitter analysis for quadrature sampling," IEEE Trans. AES-25, no. 4, July 1989, pp. 473–482.

   The performance of a adaptive clutter cancellation system is analyzed to determine the effect of time jitter. A table of resulting decorrelation for various power spectral shapes is given. Direct IF sampling provides superior performance to other methods.

3. Liu, H., A. Ghafoor, P.H. Stockmann, "A new quadrature sampling and processing approach," IEEE Trans. AES-25, no. 5, September 1989, pp. 733–748.

   A method of obtaining I and Q samples with one A/D convertor operating at a A/D conversion rate twice the bandwidth of a bandpass signal.

4. Mitchell, R.L., "Creating complex signal samples from a band-limited real signal," IEEE Trans. AES-25, no. 3, May 1989, pp. 425–427.

   A method for creating I and Q samples by sampling the IF at frequency corresponding to the bandwidth of a band-limited signal is described. The method is integrated with FIR filtering.

5. Waters, W.M., L. Jarrett, C. Lin "Processing directly sampled radar data" IEEE Int. Radar Conf., Arlington, VA, 1985, pp. 397–402.

   Description of a method of sampling IF signals and performing pulse compression prior to I,Q separation.

20.2.5   Aliasing

The alias frequency, i.e. the Nyquist frequency, is found from, [9] eq. 5.103:

$$f_a = f_{\mp n} \pm n f_s \tag{20.4}$$

where $f_{\mp n}$ is the frequency of a sampled signal indexed by $\mp n$, where n is the integer number of units of sampling frequency and the sign corresponds to the sign of f. As an example, consider a Doppler frequency of -900 Hz sampled at 500 Hz:

$$f_a = -900_{-2} + 2 \times 500 = 100 Hz \tag{20.5}$$

The alias frequency is 100 Hz.

20.3   Filtering

20.3.1   MTI

The reader is referred to the chapter on MTI. An additional bibliographical reference is:

Farina, A., A. Protopapa, "New results on linear prediction for clutter cancellation," IEEE Trans. AES- 24, no. 3, May 1988, pp. 275–286.

Detection performance is compared to the Hsaio MTI method. The authors conclude is better in terms of implementation to use as a clutter reference the first sample of the train of samples.

20.3.2   FFT/DFT

The reader is referred to the chapter on Doppler Radar. An additional bibliographical reference is:

Mitchell, R.L., "Prefolding and zero fill in FFT processing," IEEE Trans. AES-25, no.4, July 1989, pp. 580–581.

A description of a technique to prefold data prior to performing an FFT. The method is complementary to zero fill and provides the same resolution with shorter FFTs.

20.3.3   Delay-line Processing

Delay line processing can be applied to correlation, convolution, dispersion/compression, and filtering. Figure 20.1 shows the relationship between the bandwidth of a signal and the time duration of that signal that can be processed by various delay-line technologies. The ellipse areas in the figure were used by Montgomery and Dixon, [10] to illustrate the comparison between SAW and fiber-optic delay lines.

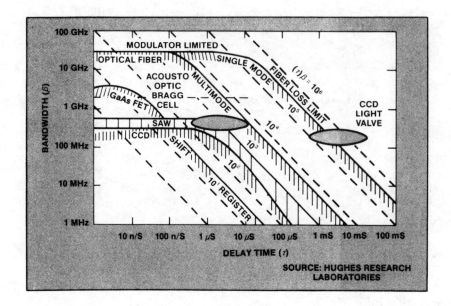

Figure 20.1: Microwave signal delay-line time and bandwidth processing capabilities. From [10].

An additional reference is:

Bowman, R., P. Gatenby, D. Switzer, "Advances in optical signal processing," Military Microwaves Conf., London, 1986, pp. 424–430.

Future applications anticipated for acousto-optic and fibre- optic delay line signal processing techniques are presented.

20.3.4   Spectral Estimation

20.3.5   Correlation

20.3.6   Pulse Compression

See 20.3.3 for time-bandwidth characteristics of delay lines.

Pulse compression sidelobes

Gating of a pulse before it has been compressed creates Fresnel sidelobes. If $TB \leq 100$ the sidelobe level will be [11]:

$$SLL = 20 \log TB + 3, \text{ dB} \qquad (20.6)$$

where SLL is relative to the peak of the compressed pulse.

Annotated bibliography

An additional bibliographical reference is:
Rohling, H., W.Plagge, "Mismatched-filter design for periodical binary phased signals," IEEE Trans. AES-25, no.6, November 1989, pp. 890–897.

A design is described where the receiver filter coefficients are selected so that all sidelobes of the cross-correlation function are zero.

20.4 Programmed Processing

20.4.1 FFT Processing

The number of operations necessary to form a spectrum of N channels in an FFT is $(N/2)\log_2 N$ complex multiplications, additions and subtractions, [12]. The total number of real operations is:

$$N_{op} = \frac{3N \log N}{\log 2} \tag{20.7}$$

The number of such operations per unit time is N_{op}/T, where T is the sample interval. T will be approximately the time extent of a range gate.

20.5 Adaptive Processing

20.6 Programmed Processing

20.6.1 General Purpose Processors

20.6.2 Special Purpose Processors

1. Allen, M.R., S.L. Katz, H. Urkowitz, "Geometric aspects of long-term noncoherent integration," IEEE Tran. AES-25, no.5, September 1989, pp. 689–700.

   An introduction and analysis of long-term integration of signals near target paths within stationary beam positions.

2. Sworder, D.D., R.G. Hutchins, "Image-enhanced tracking," IEEE Trans. AES-25, no.5, September 1989, pp. 701–710.

   A model for a processor necessary to convert a sequence of images into an estimate of target status.

3. Geideman, W.A., et al. "An all GaAs signal processing architecture," Microwave Journal, September 1987, pp. 105–126.

   Description of a spaceborne vector processor which includes device characteristics and figures of merit for signal processing.

4. Jackson, J.R., S.S. Sinor, "A modular architecture for high performance signal processing," IEEE Nat. Radar Conf., Ann Arbor, MI, 1988, pp. 188-193.

   A special purpose processor using multiple systolic arrays controlled by general purpose processing modulesis described.

5. Loppnow, D.H. "Microcomputer hardware for radar signal processing with distributed architecture," IEEE Int. Radar Conf., Arlington VA, 1985, pp. 184–189.

   A description of a special purpose computer which includes several good block diagrams.

6. Bose, N.K, (ed.) special issue on "Multidimensional signal processing," IEEE Proc., Vol. 78, no. 4, April 1990.

20.6.3   Hardwired Processors

Spearman, R., C. Spracklen, J. Miles, "The application of systolic arrays to radar signal processing," IEEE Nat. Radar Conf. Los Angeles CA, 1986, pp. 65–70.
   A description of a systolic array used to process range gated data and extract target plots.

20.6.4   Knowledge-based Processing

Vannicola, V.C., J.A. Mineo, "Applications of knowledge based systems to surveillance," IEEE Nat. Radar Conf., Ann Arbor, MI, 1988, pp. 157–164.
   A tutorial on the application of knowledge based processing of target data and control of phased array antenna, programmable signal processor, data processor and tracker of an airborne radar.

20.7   Adaptive Processing

20.7.1   Antenna Adaptive Processing

1. Adkins, L.R., et al, "Electronically variable time delays using magnetostatic wave technology," Microwave Journal, March 1986, pp. 109–120.

   The magnetostatic delay line technology is applied to the time delay steering of future phased arrays for broadband applications.

2. Wardrop, B., "Digital beamforming in radar systems, a review," Military Microwaves, London, England, 1984, pp. 319–323.

   A review of techniques including systolic arrays and FFT beamforming methods. Includes a review of experimental and commercial radars that use the methods.

3. White, W.D., "A multiple-beam architecture for sidelobe cancellers," IEEE Trans. AES-23, no. 5, September 1987, pp. 612–619.

   Mainbeam and sidelobe jammer cancellation with multiple beam adaptive cancellation.

4. Yuen, S.M., "Algorithmic, architectural, and beam pattern issues of sidelobe cancellation," IEEE Trans. AES-25, no. 4, July 1989, pp. 459–472.

   A tutorial on the subject of SLC.

20.7.2   Doppler Related Processing

20.7.3   Neural Related Processing

1. Illingworth, W.T., "Beginners guide to neural networks," IEEE AES magazine, September 1989, pp. 44–49.

   A summary of key words/phrases which includes definitions. Has a beginning tutorial and references and sources for software packages.

2. Hecht-Nielsen, R., "Neurocomputing: picking the human brain," IEEE Spectrum, March 1988, pp. 36–42.

   A listing of neurocomputers built to date and the best known neural networks. An example of recognition of an airplane regardless of its orientation is given.

3. Lau, C., B. Widrow, two special issues on Neural networks I and II IEEE Proc. vol. 78, nos. 9 and 10, September and October 1990.

Bibliography

[1] Berenz, J., B. Dunbridge, "MMIC device technology for microwave signal processing systems," Microwave Journal, April 1988, pp. 115–13.

[2] Bull, J.G., *Recent A/D convertor performance*, data provided by personal communications, October, 1991.

[3] Morchin, W.C., *Airborne Early Warning Radar*, Artech House, Norwood, MA, 1990.

[4] Mitchell, R.I., "Prefiltering cascaded stages of decimation-by-two," IEEE Trans. AES-25, no.3, May 1989, pp. 422–424.

[5] *The FFT: Fundamentals and Concepts, Instruction Manual 070-1754-00*, Tektronix, Inc., October, 1983.

[6] Barton, D.K., H.R. Ward, *Handbook of Radar Measurement*, Artech House, Norwood, MA, 1984.

[7] Agoston, M., R. Henricksen, "Using digitizing signal analyzers for frequency domain analysis," Microwave Journal, September 1990, pp. 181–189.

[8] Hansen, V.G., "Topics in radar signal processing," Microwave Journal, March 1084.

[9] Morchin, W.C., *Airborne Early Warning Radar*, Artech House, Norwood, MA, 1990.

[10] Montegomery, J.D., F.W. Dixon, "Microwave fiber optics forecast," Microwave Journal, April, 1985, pp. 44–58.

[11] Andersen Laboratories, "Pulse expansion/compression IF subsystems for radar," *Handbook of Acoustic Signal Processing*, Vol. 3, circa 1990.

[12] Bergland, G.D., "A guided tour of the fast Fourier transform," IEEE Spectrum, July 1969, pp. 41–52.

Notes

Notes

Chapter 21

Signals and Noise

21.1 Introduction

In this chapter we give data on random and spurious noise levels to be expected in the radar receiver and signal sources used in the transmitter and receiver. We give procedures for determining system temperature and noise figure including effective antenna temperature with the effects of ground and sky. Thermal noise figure of low-noise amplifiers is included. Equations with which to predict the spurious responses due to non-linear effects in amplifiers, mixers and phase detectors are presented. Typical examples of the frequency instability noise of carrier sources are tabulated. And spurious levels due to I-Q imbalance are addressed. Several of the data tabulations refer to an extensive list of references that are in themselves tabulated separately from the bibliography.

21.2 System Temperature and Noise Figure

The noise power that competes with signal detection due to various sources of noise, external and internal to the radar, is:

$$N = kT_s B \qquad (21.1)$$

where $k = 1.38 \times 10^{-23} J/K$ is Boltzmann's constant, T_s is the system temperature, and B is the signal bandwidth. The system temperature referred to the antenna terminal, [1], is:

$$T_s = T_a + T_l(L_r - 1) + L_r T_r \qquad (21.2)$$

where

T_a is the apparent temperature of the antenna due to its internal dissipative loss and its view of the external noise sources.

T_l is the temperature of the transmission line between the antenna and the receiver.

L_r is the dissipative loss of the transmission line

T_r is the effective temperature of the receiver

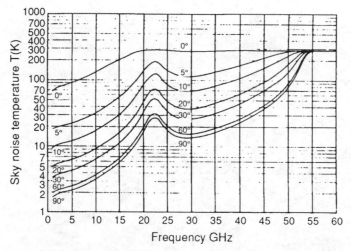

Figure 21.1: Clear-air brightness temperature for an average atmosphere from 1 to 60 GHz. The curves shown are for elevation angles, degrees, starting at the top: 0, 5, 10, 20, 30, 60, and 90. Water vapor content is $7.5g/m^3$. From Flock and Smith, [2]

The term T_r is discussed in Section 21.3.

21.2.1   Antenna Temperature

The antenna temperature is:

$$T_a = \frac{T_a'}{L_a} + T_{ta}(1 - \frac{1}{L_a})$$ (21.3)

where L_a is the antenna internal dissipative loss, $L_a > 1$, T_{ta} is the antenna thermal temperature, and T_a' is:

$$T_a' = \sum_i \alpha_i T_i$$ (21.4)

where α_i is the ith fractional angular volume that views a temperature, $T_i \cdot \sum_i \alpha_i = 1$.

Sky temperature

For skyward viewing $T_i = T_{\text{sky}}$. Background temperature values, T_b, which includes the noise contributions and losses of the earth's atmospheric gases, primarily water vapor and oxygen, are shown in Figure 21.1.

Propagation losses due to other causes such as rain or clouds or even a radome will change the brightness temperature:

$$T_{\text{sky}} = \frac{T_b}{L_p} + T_p(1 - \frac{1}{L_p})$$ (21.5)

where L_p is the one-way propagation loss, a numerical value > 1, and T_p is the average temperature of the propagating medium, rain, clouds, etc. See Chapter 15 for specific attenuation values to determine L_p.

The noise temperature of natural noise at the earth's surface is shown in Figure 21.2. These natural noise values will add to T_{sky} by adding T'_{sky} as determined in a manner similar to equation 21.5, replacing T_b with $10^{F_a/10}$, where F_a is obtained from Figure 21.2. Any non-Gaussian probability density function of a natural noise source should be considered.

Ground temperature

The nominal ground temperature is assumed to be 290 K. However, the emissivity, ϵ, will reduce the ground temperature. The emissivity depends upon, the type of surface, surface roughness, moisture content, and radar wavelength. King, [4] has shown curves of emissivity as a function of angle of incidence for K_u band that can be approximated by:

$$\epsilon = a \exp(-b\theta) \tag{21.6}$$

where θ is the angle of incidence to the earth measured from the vertical and the constants a and b are given in Table 21.1.

Table 21.1: Constants for approximating emissivity in equation 21.6.

Surface condition	a	b
Water	0.6146	0.0161
Specular moist soil	0.929	0.0136
Dry soil on rock	1.0167	0.00584

For rough soil, evaluated at the same frequency, his data showed $\epsilon = 0.99$ to be essentially independent of incidence angle. Additional data given by King showed for rough surfaces at an incidence angle of 45° that:

$$\epsilon \approx 1.12 - 0.06 \ln F_{GHz} , \text{ for } 10 \le F_{GHz} \le 75 \tag{21.7}$$

and that for frequencies from X-band to S-band that ϵ monotonically increased from 0.98 to 0.996.

Wang et al. [3], showed the ϵ dependence on soil moisture content for bare soil or lettuce vegetation at L-band at an incidence angle of 10° to be:

$$\epsilon \approx 0.97 - 0.65 \rho_w \tag{21.8}$$

where ρ_w is the soil moisture content in g/cm^3.

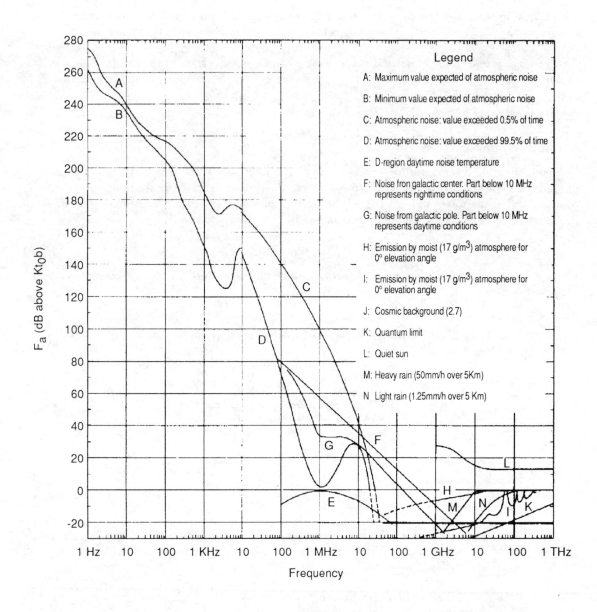

Figure 21.2: Natural noise levels in the radio spectrum. From Flock and Smith [2].

21.2.2 System Noise Figure

A common way of expressing the thermal noise is to refer to a noise figure rather than a noise temperature. The noise figure is:

$$F_n = 1 + T_s/T_o \qquad (21.9)$$

where T_o is a standard reference temperature, set at 290 K.

21.3 Device Noise Figure

21.3.1 Receiver Noise Temperature

The single response, e.g. no images or spurious responses, receiver will have an effective noise temperature of:

$$T_r = T_1 + T_2/G_1 + T_3/G_1G_2 + \ldots \qquad (21.10)$$

where T_i and G_i are the effective temperature and gain of the ith receiver amplifier stage, [6]. If there is a loss preceeding T_i, T_i is multiplied by that loss, L, and added to by the noise temperature of that loss, [6]:

$$T_{iL} = LT_i + (L-1)T_L \qquad (21.11)$$

where T_L is the temperature of the loss.

21.3.2 Receiver Noise Figure

The noise figure may be related to the effective temperature of the receiver or any amplifier by:

$$F_n = 1 + T_e/T_o \qquad (21.12)$$

where T_e is either T_r or T_i and the F_n value is specific to the device.

Several cascaded stages will have a combined noise figure of:

$$Fn = F_{n1} + \frac{F_{n2} - 1}{G_1} + \frac{F_{n3} - 1}{G_1 G_2} + \cdots \qquad (21.13)$$

where $i = 1, 2, \ldots$ refers to the stage noise figure, F_{ni} and gain, G_i. A graphical solution for two cascaded amplifiers is shown in Figure 21.3.

21.3.3 Noise Figure of a System with Feedback

Some receivers use feedback to achieve feedthrough nulling such as CW radars for cancelling unwanted leakage from the transmitter. Fahey, [7], derives the expression:

$$F_n = F_{n1} + (F_{n2} - 1)\frac{\mid G_2 \mid^2}{1 + \mid 1 - G_1 G_2 \mid^2 ((B_1/B_2) - 1)} \qquad (21.14)$$

where F_{ni}, G_i, and B_i are the noise figure, gain and bandwidths of the feed forward and feedback paths; $i = 2$ for feedback and $i = 1$ for feed forward. For large loop gain; $\mid G_1 G_2 \mid \gg 1$. For narrowband feedback; $B_1/B_2 \gg 1$. For wideband feedback; $B_1 = B_2$. For little or no feedback, $G_2 \rightarrow 0$ or $B_2 \rightarrow 0$ and $F_n = F_{n1}$.

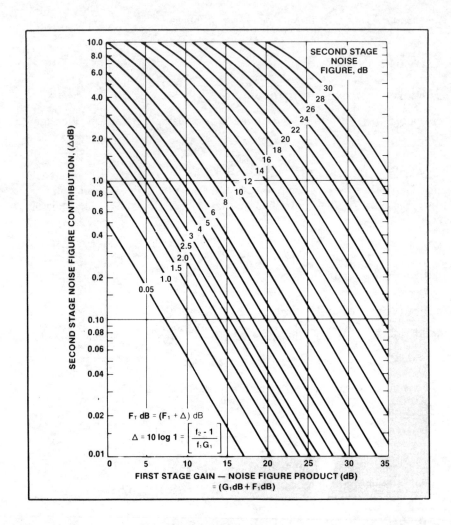

Figure 21.3: Graphical solution for noise figure of two cascaded amplifiers. From Russell [5].

21.3.4 Noise Figure of Low-Noise Amplifiers

Representative values of noise figure for low-noise amplifiers common for use in the front end of radar receivers are shown in Table 21.2. (References for Table 21.2 are given in Table 21.3.) A plot of these data is shown in Figure 21.4. Five years ago we used:

$$F_n \approx 0.7 + .58\lambda^{-0.45}, \quad \text{dB} \tag{21.15}$$

to approximate amplifier noise figure. Technology improvements are improving noise figure rapidly. Today we would use -0.25 for the exponent of wavelength in the approximation. The gain of such amplifiers generally varies inversely with bandwidth and frequency of operation. The gain typically is in the range of 10 to 40 dB, with the lower values occurring at K_u band and the higher values at S-band.

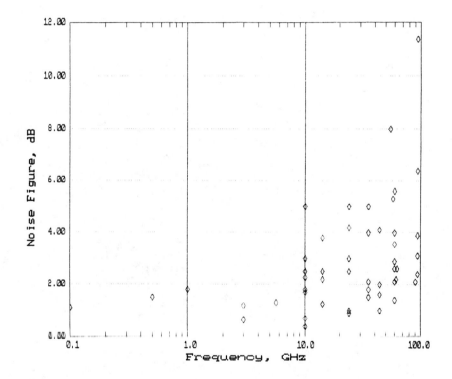

Figure 21.4: Noise figure of low-noise amplifiers.

21.4   Signal Sources

21.4.1   Phase Noise

Random phase modulation of the carrier produced by signal sources causes carrier sidebands whose amplitudes vary as a function of frequency offset from the carrier frequency. It is often referred to as single sideband noise and is the predominant factor of short term stability. Table 21.4 summarizes phase noise for various devices and frequencies. Table 21.5 summarizes the reference sources for the data. Phase noise is commonly expressed as the ratio of sideband noise power per 1 Hz bandwidth to the carrier power, given in dBc (dB below carrier). Table 21.4 uses the inverse to avoid the negative sign.

Table 21.2: Noise figure of low-noise amplifiers.

Band	Frequency (GHz)	Noise figure (dB)	Device	Reference Table 21.3
VHF	0.1	1.1	GaAs FET	(12)
UHF	0.5	1.5	GaAs FET	(12)
L-band	1	1.8	GaAs FET	(12)
S-band	3	1.2	GaAs FET	(1)
	3	0.65	MESFET	(2)
C-band	5.6	1.3	GaAs FET	(1)
X-band	10	1.8	GaAs FET	(1)
	10	1.8	MESFET	(10)
	10	2.5	MESFET	(2)
	10	0.4	Transistor	(3)
	10	0.7	Transistor	(3)
	10	3	HEMT	(2)
	10	1.7	MESFET	(2)
	10	2.26	GaAs FET	(4)
K_u-band	14	2.2	MESFET	(2)
	14	3.8	MESFET	(2)
	14	2.5	GaAs FET	(1)
	14	1.25	HEMTs	(5)
K-band	24	3	GaAs FET	(1)
	24	3	GaAs FET	(11)
	24	4.2	HEMT	(2)
	24	5	HEMT	(2)
	24	2.5	?	(6)
	24	1	Transistor	(3)
	24	1	Transistor	(3)
	24	0.9	Transistor	(3)

(Table continued on next page)

(Table 21.2 con't.)

Band	Frequency (GHz)	Noise figure (dB)	Device	Reference Table 21.3
K$_a$-band				
	35	4	MESFET	(2)
	35	2.1	HEMTs	(5)
	35	1.5	Transistor	(3)
	35	1.8	Transistor	(3)
	35	1.8	Transistor	(3)
MMW				
	44	2	HEMTs	(7)
	44	2	HEMTs	(7)
	44	1.6	HEMTs	(7)
	44	1	HEMTs	(7)
	60	2.9	HEMTs	(7)
	60	3.55	HEMTs	(5)
	60	4	Transistor	(3)
	60	2.9	HEMTs	(7)
	60	2.1	HEMTs	(7)
	60	1.4	HEMTs	(7)
	62	2.2	Transistor	(3)
	63	2.6	?	(8)
	90	2.1	?	(8)
	94	6.4	HEMTs	(7)
	94	3.9	HEMTs	(7)
	94	3.1	HEMTs	(7)
	94	2.4	HEMTs	(7)

Devices: GaAs FET is gallium arsenide field effect transistor.

HEMT is high electron mobility transistor.

MESFET is metal-semiconductor field-effect transistor, sometimes called Schottky barrier FET.

Transistor is not further defined in the reference.

? indicates not defined.

Table 21.3: Low-noise amplifier references used for Table 21.2.

(1)   Miteq, Inc., Microwave Journal, 1988.

(2)   Yeun et al., "Application of HEMT devices to MMICs," Microwave Journal, August, 1988, pp. 878–104.

(3)   Bierman, H., "Transistors stride to mm-wave performance," Microwave Journal, August 1987, p. 30.

(4)   Heston, D.D., R.E. Lehmann, "Monolithic breadboard: Key to radical MMIC size reduction," Microwave Journal, April, 1990, pp. 283–286.

(5)   Smith, A.W., "HEMTs–Low noise and power for 1 to 100 GHz," Applied Microwaves, May 1989, pp. 63–72.

(6)   Watkins, E., et al. "Low noise 20 GHz receiver front end," Microwave Journal, May 1986.

(7)   Stiglitz, M. "1989 MILCOM conference," Microwave Journal, February, 1990, pp. 28–41.

(8)   Bierman, H., "Mm-wave devices and subsystems meet military/space demands," Microwave Journal, July, 1989, pp. 26–39.

(9)   Bierman, H., "Improved devices and circuitry reduce front-end noise levels," Microwave Journal, August, 1988 (9), pp. 32–50.

(10)   Rohde, U.L., et al. "Accurate noise simulation of microwave amplifiers using CAD," Microwave Journal, December, 1988, pp. 130–134.

(11)   Hughes Aircraft Co., "EHF low noise amplifiers," Microwave Journal, Feb., 1984, p. 167.

(12)   Miteq, Inc., "Wideband, ultra-low-noise GaAs FET amplifiers outperform Bipolar," Microwave Journal, July 1987.

Table 21.4: Phase noise of signal sources. Values are carrier to noise ratio, dB, in 1 Hz bandwidth.

Device	Frequency							Ref.
	1	10	100 Hz	1	10	100	1000 kHz	
Fundamental and low-frequency sources								
synthesizer, 5 MHz	110	125	140	145	150	150		(1)
crystal oscillator, 10 MHz	100	130	150	157	159	159		(2)
PLL crystal, 100 MHz			105	130	150	158	158	(3)
PLL crystal, 100 MHz		100	140	152	173	173	172	(4)
Synthesizer, 100 MHz		70	62	85	105	130	152	(5)
UHF band								
PLL crystal		83	90	99				(6)
PLL SAW oscillator		90	108	118	133	145	145	(7)
PLL SAW oscillator		68	105	132	154	168	172	(8)
PLL SAW oscillator		75	107	132	153	168	170	(9)
L−band								
VCO Synthesizer (multiloop)		95	99	99	110	135		(10)
PLL oscillator		100	110	114	114	122		(11)
PLL oscillator	0	31	62	90	122	150		(12)
VCO (0.5 GHz tuning range)				70	127			(13)
Cavity stabilized oscillator	92	100	105	130				(10)
Cavity stabilized oscillator		80	105	130	124	145		(3)
S−band								
Cavity stabilized oscillator			74	85	85	115	130	
Cavity stabilized oscillator			52	83	112	138	157	(16)
DRO						125		(14)
PLL oscillator		80	100	120	140			(15)
YIG tuned oscillator				87	98	122	137	(17)
C−band								
PLL oscillator			92	108	113	113	130	(11)
PLL DRO			62	92	102	110	132	(18)
PLL DRO		5	40	72	108	130	142	(19)
Multiloop synthesizer			80	90	90	110	130	(10)
Cavity stabilized oscillator			80	99	99	120	138	(10)

(Continued on following page)

(Table 21.4 cont'd)

Device	1	10	100 Hz	1	10	100	1000 kHz	Ref.
X−band								
Cavity stabilized oscillator			63	88	112	107	127	
Cavity stabilized klystron			72	109	140	163		(22)
PLL oscillator			90	100	110	115	122	(11)
DRO					90	115		(20)
DRO		15	44	74	104	131	138	(21)
K$_u$ band								
Cavity stabilized oscillator			60	84	108	103	123	(3)
Cavity stabilized oscillator		40	70	74	87	108	110	(23)
DRO					85	110		(20)
DRO		40	70	82	90	109	113	(23)
DRO		28	63	92	117	117	135	(24)
PLL oscillator	39	52	60	80	91	100		(25)
PLL oscillator			83	85	90	94	94	(26)
Synthesizer	40	54	60	75	88	100		(25)
K−band								
DRO						105		(20)
DRO	0	35	65	80	95	104	115	(27)
PLL oscillator			80	90	110	114	120	(11)
PLL oscillator			77	87	87	107	126	(28)
K$_a$ band								
Cavity stabilized oscillator			132					(29)
DRO				46	78	93		(30)
Millimeter-wave								
PLL SAW oscillator (94 GHz)					55	90		(31)

Table 21.5: References noted in Table 21.4.

Reference number	Author or company, date
(1)	Pentek Inc., Microwave Journal, September 1987
(2)	Ball, Effratom Div., Microwave Journal, November 1990
(3)	Frequency-West, 1985
(4)	Driscoll, Microwave Journal, June 1990
(5)	Warwick, S., Microwave Journal, August 1987
(6)	Stirling, Microwave Journal, July 1990
(7)	Sciteg Electronics, Microwave Journal, October 1986
(8)	Sawtek Inc.,Microwave Journal, February 1989
(9)	O'Shea et al, Microwaves and RF, October 1988
(10)	Miteq, Microwave Journal, October 1987
(11)	Gamma Microwave, Microwave Journal, June 1988
(12)	Olsen and Ravid, Microwave Journal, September 1991
(13)	Z-Comm., Microwave Journal, October 1987
(14)	Bierman, Microwave Journal, October 1987
(15)	Browne, Microwaves and RF, June 1990
(16)	Huckleberry, Microwave Journal, May 1986
(17)	Grande, Microwaves and RF, September 1990
(18)	Frequency-West, 1986
(19)	Bierman, Microwave Journal, August 1988
(20)	Bierman, Microwave Journal, October 1987
(21)	MACOM, Microwave Journal, September 1987
(22)	Varian, Microwave Journal, 1986
(23)	Kumar, Microwave Journal, July 1988
(24)	Josefsberg, Microwave Journal, May 1991
(25)	Bomford, Microwave Journal, November 1990
(26)	Ohira, et al, IEEE MTT, April 1989
(27)	Gamma Microwave, Microwave Journal, January 1988
(28)	Microtech Inc., Microwaves and RF, April 1989
(29)	Strangeway, et al, Microwave Journal, July 1988
(30)	Khanna & Topacio, Microwave Journal, July 1989
(31)	Microwave Journal, July 1985

The reader is referred to Goldman, [8], for analysis and design guidance concerning signal sources and their effect on radar performance, and to Faulkner, [9], for an introduction to measurement of phase noise. Rutman and Walls [10] give a tutorial on frequency stability.

The frequency, f_n, at which phase noise begins to increase above the noise floor is a function of the noise bandwidth, B.

$$f_n \approx B/2 \approx 1/2\tau \tag{21.16}$$

where τ is the time delay used in the oscillator. As τ is made larger f_n will be moved towards the oscillator frequency. However the bandwidth is reduced which reduces the tuning range of the oscillator. One example set of data, [11], showed the phase noise for a SAW oscillator was $10\log(\Delta F^2 \times 10^{-22})$, where ΔF is the tuning range, for a 1 kHz offset from the carrier.

21.4.2   Spurious-Response Suppression in Double-Balanced Mixers

Harmonic mixing between two frequencies using a nonlinear element theoretically produces an infinite number of frequencies satisfying the condition:

$$f_I = |\pm nf_L \pm mf_R| \tag{21.17}$$

where f_L may be a local oscillator, f_R may be an rf signal, and n and m are any integer. Normally $n = m = 1$ produce the desired product, and other integer values produce spurious responses. An expression for determining the level of a spurious response of a double-balanced mixer relative to the carrier is presented by [12]. Such spurious responses are of concern when designing or using phase-locked loops, synthesizers as well as broadband receivers.

The results of measurements performed by Regev on double-balanced mixers made by two different suppliers are shown in Figure 21.5 and Table 21.6. The spurious response levels shown are relative to the rf signal carrier power, -5dBm. The spurious response levels vary with the rf signal power in a manner dependent upon the order, m, of the rf signal:

$$S_{mn} \propto P_{rf}^{m-1} \tag{21.18}$$

where S_{mn} is the spurious level and P_{rf} is the rf signal power.

We note from the data in the figure that odd harmonics of the local oscillator have higher spurious responses than do the even harmonics when mixed with odd harmonics of the rf signal. All harmonics of the local oscillator result in an approximate constant spurious level when mixed with even harmonics of the rf signal.

21.4.3   Intermodulation Products of Amplifiers

The intermodulation intercept is a measure of the linearity of an amplifier. It is determined by measuring the intermodulation ratio at an output level when signal and interference input levels are the same. The second and third order products vary as illustrated in Figure 21.6.

The product order is $m+n$. The ratio between the fundamental and the intermodulation product is:

$$\begin{aligned}
S_2(dBc) &= IP_2 - P_{out} \\
S_3(dBc) &= 2(IP_3 - P_{out})
\end{aligned} \tag{21.19}$$

where IP_i, expressed in dB units, is the ith intercept point and P_{out} is the amplifier output power, expressed in dB units.

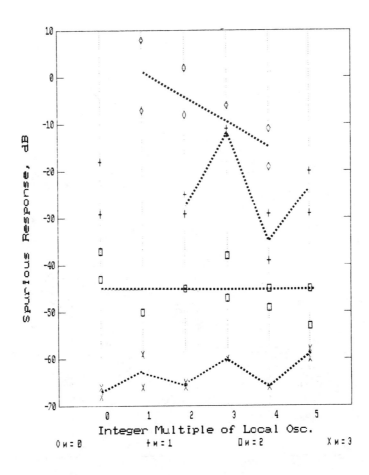

Figure 21.5: Spurious response level measurements on double-balanced mixers. After Regev [12].

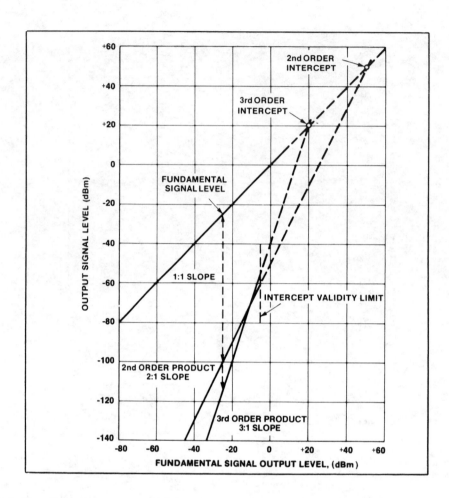

Figure 21.6: Intercept diagram. From [5].

Table 21.6: Spurious response level measurements relative to -5 dBm carrier on double balanced mixers. After Regev [12].

	Slope dependence on rf input power			
	−1	0	1	2
	Values of m (mult. of rf freq.)			
n (mult. of IF)	0	1	2	3
0		−29	−37	−66
		−18	−43	−68
1	−7		−50	−66
	8			−59
2	2	−29	−45	−66
	−8	−25		−65
3	−6	−11	−47	−60
	−6	−12	−38	−60
4	−11	−29	−45	−66
	−19	−39	−49	
5		−29	−53	−60
		−20	−45	−58

The term *intercept validity limit* in Figure 21.6 refers to the upper limit of small signal amplifier operation. Above some output level which is below the 1 dB compression point the amplifer operates in a large signal region where bias levels shift as a function of signal strength. The intermodulation products depart from straight line output slopes above this point.

Hansen predicts, [13], the third harmonic level caused by nonlinearity that has odd symmetry for positive and negative inputs. His results show the third harmonic level is:

$$S_3(dBc) = -30 + 17.73 \log G_c \qquad (21.20)$$

where G_c, in dB, is the amplifier gain compression. The results are also applicable to phase detectors.

21.4.4 Spurious Levels Due to I-Q Imbalance

Following the methods of [13], [14] and [15], the spurious level caused by amplitude and phase differences between quadrature channels, either in transmission, reception, or A/D conversion will be:

$$S(dBc) = 10 \log[(\frac{A_{dB}}{2 \times 8.62})^2 + \sin^2(\Phi/2)] \qquad (21.21)$$

where A_{dB} is the dB value for gain difference between the I and Q channels and Φ is the phase difference. A plot of equation 21.21 is shown in Figures 21.7 and 21.8. The figure with the logarithmic scales is used to show more clearly the amplitude and phase values for low spurious levels. The figure with the linear scales may be used to determine values intermediate between the limits of the curves more easily for the higher spurious levels. The linear scaled figure can be used to approximate values for the low valued spurious levels by noting: amplitude and phase values are divided by 10 for each 20 dB decrease in spurious level, and are divided by $\sqrt{10}$ for each 10 dB decrease in spurious level.

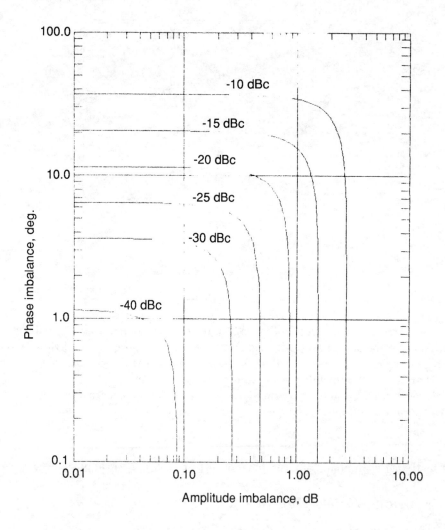

Figure 21.7: Combinations of amplitude and phase imbalance in quadrature processors for various spurious response levels – logarithmic scaling.

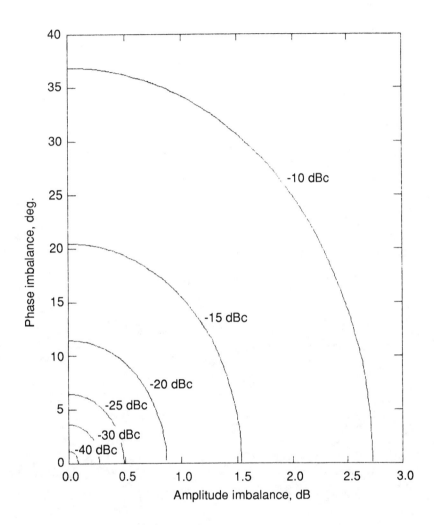

Figure 21.8: Combinations of amplitude and phase imbalance in quadrature processors for various spurious response levels – linear scaling.

Bibliography

[1] Blake, L.V., *Radar Range-Performance Analysis*, Artech House, Norwood, MA, 1986.

[2] Flock, W.L., E.K. Smith, "Natural radio noise –A mini-review," IEEE Trans. AP-32, no. 7, July, 1984, pp. 762–767.

[3] Wang, J.R. et al., "The effects of soil moisture, surface roughness, and vegetation on L-band emission and backscatter," IEEE Trans. GE-25, no. 6, November 1987, pp. 825–833.

[4] King, D.D., "Passive Detection," Chapter 39 of Radar Engineers Handbook, M.I.Skolnik, ed., McGraw-Hill, New York, 1970.

[5] Russell, A., "Amplifier application notes," Microwave Journal, October, 1982, pp. 130–136.

[6] Mumford, W.W., E.H. Scheibe, *Noise Performance Factors in Communication Systems*, Artech House, Norwood, MA, 1968.

[7] Fahey, S.F., "Noise factor of a system with feedback," Microwave Journal, September, 1989, pp. 228–236.

[8] Goldman, S.J., *Phase Noise Analysis in Radar Systems Using Personal Computers*, John Wiley and Sons, New York, 1989.

[9] Faulkner, T.R., "Residual phase noise measurement," Microwave Journal Reference Issue, September 1989, pp. 135–143.

[10] Rutman, J., F.L. Walls, "Characterization of frequency stability in precision frequency sources," IEEE Proc. vol. 79, no. 7, July 1991, pp. 952–960.

[11] Anderson Laboratories, *Handbook of Acoustic Signal Processing*, vol. 4, 1989.

[12] Regev, D., "Characterization of spurious- response suppression in double-balanced mixers," IEEE Trans. MTT-38, no. 2, February 1990, pp. 123–128.

[13] Hansen, V. Gregers, "Topics in radar signal processing," Microwave Journal, March 1984, pp. 24–46.

[14] Sinsky, A.I., P. Wang, "Error analysis of a quadrature coherent detector processor," IEEE Trans. AES-10, no. 6, November 1974, pp. 880–883.

[15] Monzingo, R.A., S.P.Au, "Evaluation of image response signal power resulting from I-Q channel imbalance," IEEE Trans. AES-23, no. 2, March 1987, pp. 285–287.

Notes

Notes

Chapter 22

Synthetic Aperture Radar

22.1 Introduction

We summarize synthetic aperture radar (SAR) technology with tables and listings of system defining equations. In addition to the references used for those data, the reader is referred to [1], [2], and [3] for discussions of the theory, applications, implementations, and design constraints of SAR.

22.2 Basic Side-Looking SAR Equations

Basic side-looking synthetic aperture radar (SAR) equations are summarized in Table 22.1. A squint angle, the angle relative to the orthogongal to the SAR velocity vector, is allowed by the use of the angle, θ_v, the angle from the velocity vector to the antenna beam line of sight. A flat earth and level flight is assumed.

Table 22.1: Summary.

Parameter	Equation	Reference
Power Aperture	$P_{avg}A_e^2 = \dfrac{8\pi R_o^3 kT_s LV\lambda(S/N)sin\theta_v}{\sigma_o \Delta R_r}$	[6]
Azimuth Resolution (units of length), Unfocused Array	$r_y = \sqrt{\lambda R_o/2}$	[6]
Azimuth Resolution (units of length), Focused Array	$r_y = l/2$	[6]
Resolution Perpendicular to Flight Path	$r_x = c\tau_c \cos\theta_v/2\sin\theta_g$	[6]
PRF Limits	$2VK_a/l \leq f_r \leq c/2R_{max}$	[6]
Antenna Height	$h = \dfrac{2K_e R_o \lambda f_r}{c\tan\theta_d}$	[8]

431

Table 22.1, (Continued)

Parameter	Equation	Reference
Swath width	$S_x = \dfrac{R_o \lambda}{l \sin \theta_g}$	[6]
Swath length	$S_y = R_o \lambda / l$	[6]
Number of range gates within beamwidth	$N_R = \dfrac{2R\lambda}{lc\tau_c \tan \theta_g}$	
Number of azimuth samples per range gate	$N_D = \dfrac{4\lambda R_o}{r_y^2}$	[6]
Total number of samples	$N = \dfrac{4R_o^2 \lambda^2}{r_y^2 r_x^2 l \sin \theta_g}$	[6]
Coherent integration time	$T_{coh} = \dfrac{\lambda R_o}{2 r_y V \sin \theta_v}$	[6]
Allowed cross-track acceleration for focused array ($\pi/2$ phase error at edge of synthetic aperture)	$a = \dfrac{4V^2 r_y}{\lambda R_o^2}$	[5]
Aperture time (focused)	$T_a = S_y/V = R_o \lambda / lV$	
(unfocused)	$T_a = \sqrt{\dfrac{\lambda R_o}{V \sin \theta_g}}$	[4]
Depth of cross range focus	$DOF_x \approx \dfrac{2\lambda R_o^2}{(V T_a \sin \theta_v)^2}$	[4]
	$DOF_x \approx \dfrac{2l^2}{\lambda \sin^2 \theta_v}$	
Image formation condition ($\theta_v = 90°$)	$r_y^2 r_x \geq \lambda^2 R_o / 16$	[7]

R_o is radar range from center of synthetic aperture to a point scatterer
$R_{\max}$ is the maximum 3 dB antenna illuminated range
T_s is radar system effective temperature
L ares the radar system losses
V is the radar platform velocity
(S/N) is the signal-to-noise ratio required for detection
θ_v is the angle between the radar platform velocity vector and the antenna mainbeam
σ_o is the reflectivity of the target surface
ΔR_r is the radar resolution along the ground in alignment with the beam pointing direction
f_r is the PRF
l is the antenna length
w is the antenna height

K_a is an aperture factor for the azimuth antenna dimension. It is:

$\quad K_a \quad = 0.885$ for uniform aperture, [8]

$\qquad\qquad = 1$, [6]

$\qquad\qquad = 1.186$ for cosine weighted aperture, [8]

K_e is an aperture factor for the elevation antenna dimension. It is 0.885 for uniform illumination and 1.856 for a cosine aperture illumination weighting

τ_c is the compressed pulse width

θ_g is the grazing angle

θ_d is the depression angle from a horizontal plane through the SAR antenna phase center.

22.3   Inverse Synthetic Aperture Radar (ISAR)

Cross-range resolution is [7]:

$$\delta_x = \lambda/2\omega T_{coh} \tag{22.1}$$

where ω is the object rotation rate, coplanar with the object down-range dimension and the radar line-of-sight. Also:

$$\delta_x = \lambda/2\Delta\alpha \tag{22.2}$$

where $\Delta\alpha$, radians, is the object rotation angle that occurs during T_{coh}.

The down-range resolution is [7]:

$$\delta_r = c/2B \tag{22.3}$$

where B is the radar signal bandwidth.

For target motion to be constrained to the resolution cell:

$$\begin{array}{l} T_{coh} < \omega^{-1}\sqrt{\lambda/L_r} \quad \text{and} \\ T_{coh} < c/2B\omega L_x \end{array} \tag{22.4}$$

where L_r is the down-range object dimension and L_x is the object cross-range dimension. x and r are coplanar with R_o and ω is constant.

Also:

$L_r < 4\delta_x^2$

$L_x < 4\delta_x\delta_r/\lambda$

The unambiguous interval in cross range is [2]:

$$L_{xu} = \lambda/2\Delta\alpha \tag{22.5}$$

The number of measurement samples required is [9]:

$$N_m = L_x/\delta_x T_{coh} \tag{22.6}$$

and the sampling rate is:

$$f_s = N_m/T_{coh} \tag{22.7}$$

22.4   SAR with a Rotating Antenna

In this case with a rotating SAR antenna the illuminated swath is a circular ring.

Angular resolution for large time-bandwidth product is [10]:

$$\alpha_{az} = \lambda/5L_b \sin(\theta/2), \text{ radians} \tag{22.8}$$

where L_b is the radius of antenna rotation, from center of rotation to antenna phase center, and θ is the real beamwidth of the antenna.

22.5   Bistatic SAR

The bistatic Doppler bandwidth for a target to be imaged is [11]:

$$\Delta f_D = \frac{r_e \cos(\beta/2)}{\lambda} \left(\frac{V_T}{R_T} \sin\theta_T + \frac{V_R}{R_R} \sin\theta_R \right) \tag{22.9}$$

where:

r_e is the separation of two points on a range ellipsoid to be resolved,
β is the bistatic angle,
V_T is the transmitter ground speed,
V_R is the receiver ground speed,
θ_T is the transmitter line-of-sight angle relative to transmitter platform velocity vector, and
θ_R is the receiver line-of-sight angle relative to the receiver platform velocity vector.

And:

$$T_{coh} = K_s/\Delta f_D \tag{22.10}$$

where K_s is a factor to account for the effects of weighting used to reduce sidelobes, $K_s \approx 1.2$ for -30 dB Doppler sidelobes.

The stability of the clock used to determine when the transmitter pulse is sent must be, [11]:

$$s_c = r/ct_F \tag{22.11}$$

where r is the required range accuracy, t_F is the time over which a SAR scene is measured.

The required receiver frequency stability when phase locked to the transmitter signal must be [11]:

$$s_r = \frac{\epsilon_\Phi c}{f_o(R_T + R_R - R_B)} \tag{22.12}$$

where ϵ_Φ is the allowable phase error, f_o is the bistatic radar frequency, and R_B is the receiver to transmitter baseline distance.

If instead of a phased-locked receiver, coherency is obtained by stable free-running oscillators in both the transmitter and receiver, the oscillator stability in each must be, [11]:

$$s_o = \epsilon_\phi/f_o T_{coh} \tag{22.13}$$

where ϵ_ϕ is the allowable quadratic phase error. A summary of typical values of the stability requirements is shown in Table 22.2.

Table 22.2: Typical stability required for bistatic SAR.

Stability requirement	Typical value
Ranging	10^{-12}
Phase locked local oscillator	10^{-9}
Free-running local oscillator	10^{-11}

22.6   SAR Processing

Otten, [13], presents equations for predicting the number of FFT multiplies that are necessary to autofocus a SAR by two different means; a map drift method and a shift and add method used in optical astronomy.

Table 22.3: SAR digital processing requirements for different implementations

SAR Implementation	Arithmetic rate and bulk memory	Reference
Correlator		
Arithmetic	$N_D N_R f_r$	[12]
	$S_y N_R f_r / r_y$	[11]
Memory	$2 k_b N_D N_R$	[12]
	$f_r S_y N_R / r_y$	[11]
Prefilter plus correlator		
Arithmetic	$N_D f_r + (K_{OS} K_S N_D^2 N_R / T_{coh})$	[12]
	$(2 S_y^2 N_R)/(r_y^2 T_{coh})$	[11]
Memory	$2 k_b N_D N_R$	[12]
	$k_b R_o \lambda N_R / r_y^2$	[11]
Prefilter plus two stage correlator		
Arithmetic	$N_R f_r + (K_{OS} K_S N_R N_D^{3/2} / T_{coh})$	[12]
	$(4 S_y^{3/2} N_R)/(r_y^{3/2} T_{coh})$	[11]
Memory	$2 k_b K_{OS} K_S N_R [M + (M+3) N_D / 2M]$	[12]
FFT after prefilter		
Arithmetic	$N_R f_r + (K_{OS} K_S N_D N_R \log_2)$	
	$(K_{OS} K_S N_{ipp} / 2 T_{coh})$	[12]
	$(2 S_y \log_2(2 S_y / r_y) N_R)/(r_y T_{coh})$	[11]
	$V N_R N_D \log_2 N_D / r_y$	[6]
Memory	$2 k_b K_{OS} K_S N_D N_R$	[12]

Note:
f_r is the PRF
k_b is the number of sample amplitude bits
K_{OS} is the prefilter oversample factor, ≈ 1.7
K_S is synthetic array weighting constant, ≈ 1.2
M is number of first-stage correlator filters
N_{ipp} is number of intepulse periods
N_D, N_R, S_y, r_y, T_{coh}, and V are defined in Table 22.1.

Bibliography

[1] Skolnik, M.I., *Radar Applications*, IEEE Press, New York, 1987, papers 5.1, 5.2, and 5.4.

[2] McCandless, S.W., "SAR in space-the theory, design, engineering and application of a space-based SAR system," L.J. Cantafio, ed., Ch. 4, *Space-Based Radar Handbook*, Artech House, Norwood, MA, 1989.

[3] Cutrona, L.J., "Synthetic aperture radar," M.I. Skolnik, ed., Ch. 21, *Radar Handbook*, McGraw-Hill, New York, 1990.

[4] Morris, G.V., *Airborne Pulsed Doppler Radar*, Artech House, Norwood, MA, 1988.

[5] Wood, J.W., et al. "Distortion free SAR imagery and change detection," IEEE National Radar Conference, Ann Arbor, MI, April 20–21, 1988, pp. 95–99.

[6] Hovanessian, S.A., *Introduction to Synthetic Array and Imaging Radars*, Artech House, Norwood, MA., 1980.

[7] Ausherman, D.A., et al. "Developments in radar imaging," IEEE Trans. AES-20, no.4, July 1984, pp. 363–400.

[8] Bayma, R.W., P.A. McInnes, "Aperture size and ambiguity constraints for a synthetic aperture radar," IEEE Int. Radar Conf., Arlington, VA, 1975, pp. 499–504.

[9] Corsini, G., et al., "Radar imaging of noncooperating maneuvering aircraft," IEEE Int. Radar Conf., RADAR-90, Arlington, VA, May 7–10, 1990, pp. 563–568.

[10] Klansing, H., "Synthetic aperture radar with rotating antennas," Military Microwaves Conf., London, July 11–13, 1990, pp. 530–538.

[11] Kirk, J.C., "Bistatic SAR motion compensation," IEEE Int. Radar Conf., Arlington, VA, May 6–9, 1985, pp. 360–365.

[12] Mrstik, A.V., *RF Systems in Space, RADC-TR-83-91*, vol. 2, April, 1983, AD-A133735.

[13] Otten, M.P.G., "Comparison of SAR autofocus algorithms," Military Microwaves, London, 1990, pp. 362–367.

Notes

Notes

Chapter 23

Tracking Radar

23.1   Introduction

Radar tracking is the determination of the path of various targets and their relationship with each other and geometrical points of reference. A thorough coverage of the subject is given by Blackman, [1]. Other useful references, in addition to those used in the text, are Howard, [2], Trunk, [3], Barton, [4], and Bar-Shalom, [5]. The reader is referred to these references for coverage of the wide variety of tracking techniques and implementations available. We summarize material pertaining to angle tracking with computer controlled tracking algorithms. We exclude range and Doppler domain tracking and tracking with servo controlled pointing of motor driven antennas.

23.2   Computer Tracking of Target Position

The performance of computer tracking of targets can not be generalized so that one can make specific statements on anything other than case studies or tests. The processing of tracks is not general because it depends on the many branches in the track logic, on the evolutions of target(s) during the whole track process, on the environment and on the behavior of the radar. Design iterations and operational success depend on case trials.

23.2.1   Equations of State for Tracking

The track filter equations used to smooth the variations of radar reports and to predict the next target position reported by the radar are:

$$
\begin{aligned}
E_{xyz}(n) &= F(n) + \alpha[M(n) - F(n)] \\
E_v(n) &= E_v(n-1) + T_r E_a(n-1) + (\beta/T_r)[M(n) - F(n)] \\
E_a(n) &= E_a(n-1) + (\gamma/T_r^2)[M(n) - F(n)] \\
F(n+1) &= E_{xyz}(n) + E_v(n)T_r + E_a(n)T_r^2/2
\end{aligned}
\tag{23.1}
$$

where:

$E_{xyz}(n)$ is a position matrix in the x, y, and z coordinate system denoting the expected position of the target. It is the expected target position for the present look n.

441

$F(n)$ is a forecasted position matrix based on processing prior to the present radar look n.

$M(n)$ is the radar reported measurement of position converted to x, y, and z coordinate system,

$E_v(n)$ the estimated target velocity matrix in the x, y, and z coordinate system,

$E_a(n)$ is the estimated target acceleration matrix in the x, y, and z coordinate system,

T_r is the time between radar measurements, α, β, and γ are smoothing factors.

Some tracking systems use polar coordinates in the tracking equations and subsequently convert the tracks to the x, y, and z coordinate system for correlation with other tracks or geographic position.

The smoothing factors may be fixed or varied during the tracking process. It is usual to vary the factors. Symon [6] uses:

$$\alpha = \frac{4N - 2}{N(N + 1)}$$
$$\beta = \frac{6}{N(N + 1)} \tag{23.2}$$

where N is a quality index which increases by one each time a forecast dimension is sufficiently accurate until a prescribed limit is reached. Otherwise, N decreases by one each time until a prescribed lower limit is reached.

β and γ are normally computed using values of α that have been derived by an adaptive procedure in the tracking computer. McLane et al. [7], use:

$$\beta = 4 - 2\alpha - 4\sqrt{1 - \alpha}, \alpha \leq 1$$
$$\gamma = \beta^2 / 2\alpha \tag{23.3}$$

Trunk [3] reports:

$$\beta = \alpha^2 / (2 - \alpha) \tag{23.4}$$

for good maneuver-following capability.

23.2.2 Evolution of Target Track Status

The evolution of the target track status is normally modeled as a sequence of steps referred to as a Markov chain, [5]. Each step uses a modeled radar performance result as input to modeled tracking algorithms. Table 23.1 shows how such modeling can branch to various radar/track modes depending upon the probability or non-probability of target detection. The table shows how the radar tracking changes status from radar look i to look $i + 1$. One would also need to duplicate this table, replacing the entries with probability and non-probability of false reports.

The notation used in Table 23.1 is:

1. S is search

2. A_1 waiting for a detection at the next scan or look to initiate a track

3. A_2 waiting for a detection within the next two scans or looks to initiate a track

4. T_1 waiting for a detection at the next scan or look to continue the track

5. T_2 waiting for a detectioin at the next two scans or looks to continue the track

6. T_3 waiting for a detectioin at the next three scans or looks to continue the track

7. T_4 waiting for a detectioin at the next four scans or looks to continue the track

Table 23.1: Evolution of the state vector from scan (i) to $(i + 1)$.
After Ramstein and Georges [8].

			Scan (i+1)				
Scan (i)	$S(i+1)$	$A_2(i+1)$	$A_1(i+1)$	$T_4(i+1)$	$T_3(i+1)$	$T_2(i+1)$	$T_1(i+1)$
$S(i)$	$1 - Pd_S$	Pd_S					
$A_2(i)$			$1 - Pd_A$	Pd_A			
$A_1(i)$	$1 - Pd_A$			Pd_A			
$T_4(i)$				Pd_T	$1 - Pd_T$		
$T_3(i)$				Pd_T		$1 - Pd_T$	
$T_2(i)$				Pd_T			$1 - Pd_T$
$T_1(i)$	$1 - Pd_T$			Pd_T			

Note: Pd_S is the probability of detection in search
Pd_A is the probability of detection in acquisition
Pd_T is the probability of detection in track

23.2.3   Track Performance Prediction

Bar-Shalom et al. [9], have presented a Markov-chain based performance evaluation technique for predicting the performance of a computer controlled tracking system. Their track system depended on a cascaded detection logic:

1. **2/2 requirement.** Following a first look detection, a logic volume gate is established within which a second look detection must occur.

2. **m/n requirement. m** detections must occur within the next **n** looks if the **2/2 requirement** is met. The detections must occur within logic volume gates that have been established by the assumed target motion model.

Their results, [9], can be summarized by the following approximating functions for nearly constant velocity targets with piecewise white noise acceleration with variance of $q = 0.1$, [10]. The average time in units of number of scans to confirm a true track is:

$$\overline{T}_T \approx A \exp[(\ln P_d - B)^2/C] \tag{23.5}$$

where P_d probability of detecting the true target, and the other constants are defined in Table 23.2. The probability of false track is:

$$P_{FT} \approx DP_{fa}^E \tag{23.6}$$

where P_{fa} is the probability of false alarm and the other constants are shown in Table 23.2. The average false track length in units of number of scans is:

$$\bar{t}_F \approx F + G \ln P_{fa} + H/\ln P_{fa} \qquad (23.7)$$

where the constants are shown in Table 23.2. The logic gate volumes in two dimensions for the performance predicted with equations 23.5 through 23.7 is 400 resolution cells for the second look of the **2/2 requirement** and 150 for $m = 1$ of the **m/n requirement**.

Table 23.2: Table of constants for use in predicting track performance
using equations 23.5, 23.6, and 23.7.

Detection requirement	Constants							
m/n	A	B	C	D	E	F	G	H
1/2	1.3096	1.2126	1.7761	2E5	3	−13.29	−0.6641	−77.44
2/3	2.6336	0.6017	0.8761	1.69E7	4	−6.662	−0.3553	−41.54
2/4	2.6292	0.7279	1.2516	3.73E7	3.94	−8.881	−0.4743	−51.77
3/4	3.6381	0.4182	0.5676	1.09E9	4.98	−8.763	−0.4672	−51.29

The number of false tracks can be estimated from:

$$N_F = N_C P_{FT} \bar{t}_F \qquad (23.8)$$

where N_C is the number of radar resolution cells.

23.2.4   Track Sampling Time

The necessary time-between samples, e.g. the track revisit time is [16]:

$$T_R = \sqrt{\frac{2\phi R_{min}}{a_r}} \qquad (23.9)$$

where ϕ is the allowed angle change of the radar line- of-sight from one track sample to the other, R_{min} is the minimum sampled range of the target, and a_r is the radial acceleration of the target about its turning point. ϕ may be determined by referring to equations 23.15 and 23.16.

Van Keuk, [11], presents a technique for determining T_R with a computer used for track:

$$T_R = 0.4 \left(\frac{\sigma_{Lx} \sqrt{t_{ar}}}{a_r} \right)^{0.4} \frac{V_o^{2.4}}{1 + V_o^2/2} \qquad (23.10)$$

where σ_{Lx} is the rms position noise, t_{ar} is the correlation time of a target maneuvor, and V_o is a measure of track sharpness:

$$V_o = \frac{0.42 R_{min} \theta_{-3}}{k \sigma_{Lx}} \qquad (23.11)$$

where k is a selectable parameter which depends upon the desired probability of the target being located within the half-power beamwidth of the track beam. For a Gaussian distributed cross range target position $k = 2.3$ for a 99% probability. A target acceleration capability value may be used for a_r.

An example for a low rate target maneuver is $a_r = 1m/s^2$ and $t_{ar} = 100s$. A high rate maneuver is $a_r = 20m/s^2$ and $t_{ar} = 30s$.

23.3   Radar Report Angle Errors

Barton, [17] Ch. 8, gives a good tutorial and performance data concerning angle measurement and expected measurement accuracy obtainable with a radar. We summarize relationships useful for extending or investigating his generalized error data. Additional information on the topic can be found in Chapter 10, Measurement.

Jacovitti, [12], gives approximations for the sum and difference beams useful for investigating independent control of sum and difference beams. Ewell and Reedy give similar expressions using Gaussian beam shapes, [13]. For the sum beam [12]:

$$G_{\Sigma(y)} = \frac{\sin(3\pi y/2)}{3\pi y/2} \frac{1}{1 - (3y/2)^2} \tag{23.12}$$

where y is the off-boresight angle normalized by the half-power beamwidth of the sum beam. For the difference beam:

$$G_\Delta(y) = \frac{\eta(2/\pi)\sin(3\pi y/2)}{1 - (3y/2)^2} \tag{23.13}$$

where η is a gain factor of the difference beam relative to the sum beam. These functions and G_Δ/G_Σ are shown in Figure 23.1.

The slope of the $G_\Delta(y)/G_\Sigma(y)$ amplitude patterns is given by:

$$k_m = 3\eta \tag{23.14}$$

Practical values of η are 0.5 to about 0.8.

The expression for the rms angle error normalized by the half-power beamwidth of the sum pattern is [14]:

$$\frac{\sigma_\theta}{\theta_{-3}} = \frac{\sqrt{1 + (y\bar{k}_m)^2}}{k_m\sqrt{2n(S/N)}} \tag{23.15}$$

where

k_m is slope of the difference beam to the sum beam amplitudes,

$\bar{k}_m$ is the average of k_m from 0 to the target position,

n is the number of independent samples, each with

S/N target signal-to-noise ratio.

Following the method given by Fedele et al. [15], for a sum and difference channel phase monopulse system the error for a single nonfluctuating pulse is:

$$\frac{\sigma_\theta}{\theta_{-3}} = \frac{\sqrt{1 + (3\eta\pi y)^2}}{3\eta\sqrt{2(S/N)}} \tag{23.16}$$

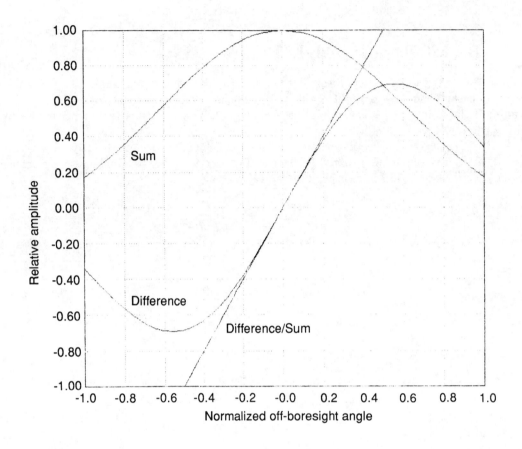

Figure 23.1: Sum and difference beam and difference/sum ratio with $\eta = 2/3$, equations 23.12 and 23.13.

Bibliography

[1] Blackman, S.S., *Multiple-Target Tracking with Radar Applications*, Artech House, Norwood, MA, 1986.

[2] Howard, D.D., "Tracking Radar," Ch. 18, M.I. Skolnik, ed., *Radar Handbook*, McGraw-Hill, New York, 1990.

[3] Trunk, G.V., "Automatic Tracking" sec. 8.3, *Radar Handbook*, M.I. Skolnik, ed., McGraw-Hill, New York, 1990.

[4] Barton, D.K., Radars–Volume 1, *Monopulse Radar*, Artech House, Norwood, MA, 1975.

[5] Bar-Shalom, Y., ed., *Multitarget-Multisensor Tracking: Advanced Applications*, Artech House, Norwood, MA.,

[6] Symon, M., "The automatic track while scan system used within the Searchwater airborne maritime surveillance," IEE Int. Radar Conf., London, 1982, pp. 254–258.

[7] McLane, P.J., P.H. Wittke, C.I., "Least mean-square-error adaptation of parameters in track-while-scan radar systems," IEEE Int. Radar Conf., Arlington, VA, 1980, pp. 451–457.

[8] Ramstein, Georges, "A new criterion to assess radar performances," IEE Int. Radar Conf., London, 1987, pp. 409–413.

[9] Bar-Shalom, Y., K. Chang, H. Shertukde, "Performance evaluation of a cascaded logic for track formation in clutter," IEEE Trans. AES-25, no. 6, November 1989, pp. 873–877.

[10] Bar-Shalom, T. Fortmann, *Tracking and Data Association*, Academic Press, 1988.

[11] van Keuk, G., "Adaptive computer controlled target tracking with a phased array radar," IEEE Int. Radar Conf., Arlington, VA, 1975, pp. 429–434.

[12] Jacovitti, G., "Performance analysis of monopulse receivers for secondary surveillancee radar," IEEE Trans. AES-19, no. 6, November 1983, pp. 884–897.

[13] Ewell, G.W., E.K. Reedy, "Multipath effects on low-angle millimeter wavelength radar tracking," IEE Int. Radar Conf., London, 1987, pp. 423–427.

[14] Sherman, S., *Monopulse Principles and Techniques*, Artech House, Norwood, MA, 1984.

[15] Fedele, G., M.Orsini, S. Strappaveccia, "Analysis of processors and post-detection algorithms for monopulse radar," IEEE Int. Radar Conf., 1985, pp. 88–94.

447

[16] Morchin, W.C., *Airborne Early Warning Radar*, Artech House, Norwood, MA, 1990.

[17] Barton, D.K., *Modern Radar System Analysis*, Artech House, Norwood, MA, 1988.

Notes

Notes

Chapter 24

Transmitters

24.1 Introduction

Transmitters and the sources of rf power, tubes and solid-state devices, are discussed in detail in several common radar books. The reader is referred to Skolnik, [1], Weil, [20], Borkowski, [2], Staprans, [21], and Smith, [4] for additional information.

Vacuum tubes remain as the primary high-power source of rf. However, in some cases vacuum tube rf power sources are being replaced by solid-state power. An example is the AN/SPS-40 radar, [5]; a nominal 250 kW UHF transmitter. The transmitter is centrally located and it consists of 112 parallel connected power amplifier modules. Other examples are the RAMP, [6], [7], and the KASTRUP, [8], radars used for air-traffic-control surveillance. A further example is the 217 MHz 770 kW space surveillance radar, [9].

Bhanji, et al. [10] describe a conceptual design of a 400 kW cw transmitter operating at 38 GHz that uses a klystron tube.

Wallington and Chrystie, [11] describe a demonstration solid-state active planar array system that accomplishes aperture amplitude taper be means of couplers and variable phase shifters arranged along line arrays of a planar array.

24.2 Comparisons of Vacuum and Solid State Electronics

24.2.1 Average Power as a Function of Frequency for Tube and Solid-State RF Sources

Figure 24.1 shows the average power capability of vacuum tube and solid-state rf power sources.

Vacuum tube power

Further information to that given above, which is specific to vacuum tube power given here. Hansen, [13], shows for the traveling-wave tube that the power capability is derated about 1/2% for every 1% of rf bandwidth above a bandwidth of 5%. Belna et al. [14] show in Figures 24.2 and 24.3 the relationship between peak and average power limits.

We see that the upper average power limit is reached at a duty factor in the 2% to 5% region for a constant peak power, after which the average power limit decreases for increasing duty factor.

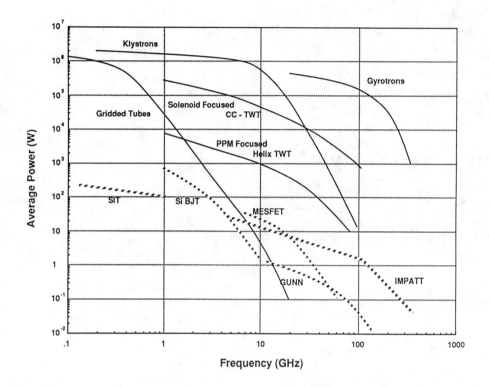

Figure 24.1: 1990 average power capability of vacuum tube and solid-state rf power source technology. From Sleger et al. [26]. See also [12].

The average power limit decreases about 0.25 dB for every 1 dB increase in duty factor, for duty factors beyond the upper average power limit.

Additional definition of the relationship between peak and average power for TWTs is shown in Figures 24.4 and 24.5. Specific information on the characteristics of millimeter-wave vacuum power tubes is shown in Table 24.1.

Table 24.1: High-power millimeter-wave tubes. After Bierman [15].

Frequency (GHz)	Peak power, W	Average power, W	Gain (dB)	Structure	Focusing	Weight (kg)	Specific weight (kg/W)
18 to 40	20	20	40	helix	ppm	1.1	0.054
27 to 40	15	15	30	helix	ppm	2.5	0.166
32 to 35	80	16	43	helix	ppm	2.5	0.156
34 to 36.5	4000	220	46	cavity	ppm	7.9	0.036
40	1000	1000	43	cavity	ppm	2.7	0.003
34.5 to 35.5	30000	3000	50	cavity	solenoid	158.8	0.053
34 to 60	50000	5000	40	cavity	solenoid	158.8	0.032
93.7 to 95.7	100	50	60	cavity	ppm	6.4	0.127
94.5 to 95.5	5000	500	46	ladder	solenoid	99.8	0.200
80 to 100	100	100	40	millitron	ppm	6.8	0.068

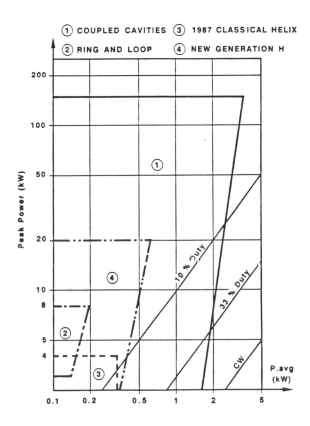

Figure 24.2: X-band TWT peak/average power limits for different available technologies. From Belan et al. [14].

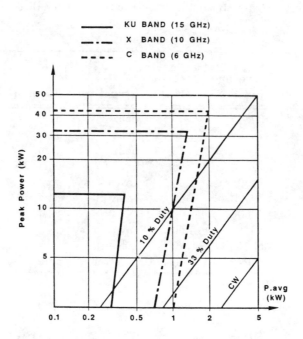

Figure 24.3: TWT peak/average power limits for grid-pulsed, pressed-helix and PPM-beam technologies. From Belan et al. [14].

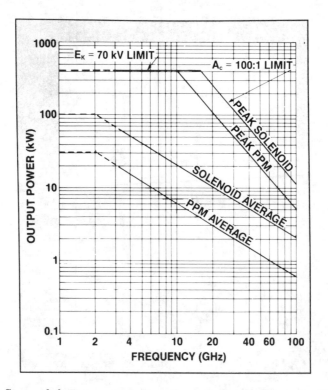

Figure 24.4: 1989 State-of-the-art output power for pulsed coupled-cavity TWTs having moderate bandwidth. From Hansen, [13].

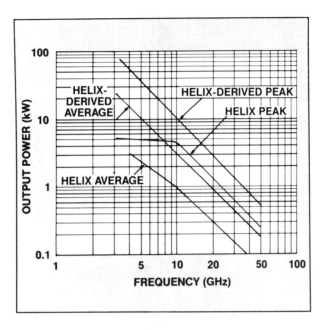

Figure 24.5: 1989 State-of-the-art output power for pulsed helix TWTs. From Hansen [13].

Solid-state power

Thermal effects cause nonlinear relationships between maximum power capability of solid state devices for various combinations of pulse width and duty factor. Such a relationship is shown in Figure 24.6.

The curves in Figure 24.6 follow the approximation:

$$P_o \approx 15.5\tau^{-0.02}df^{-0.02}, 1 \leq df < 10$$
$$P_o \approx K15.5\tau^{-0.02}df^{-0.082}, 10 \leq df \leq 50 \tag{24.1}$$

where τ is the pulse width, μs, df is the duty factor, %, , and K is a curve fitting constant, 1.16, in this example. If we can generalize, for duty factors greater than some value near 10% the average power capability applicable to radar applications is:

$$P_{avg} = KP_{max}\tau^{-a}df^b \tag{24.2}$$

where P_{max} is the device maximum power capability for low duty factor and short pulse conditions, and a is a small value such as 0.02 in this example, and b is a value near 1 such as $1 - 0.082 = 0.916$ in this example.

Rather than obtaining rf power by means of fundamental frequency oscillators or amplifiers, one can use frequency multipliers. Typically power at high frequencies with solid-state devices has been obtained in this manner. Low-loss varactor diodes are used for this purpose. Stacking

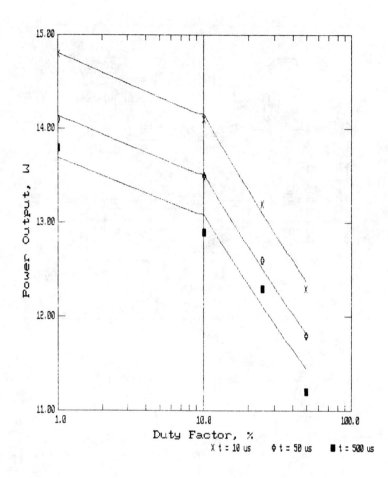

Figure 24.6: Maximum power capability of a S-band FET MMIC power amplifier. Data points after [16].

such diodes has been used to increase breakdown voltage for increased power capability. Staecker, [18], demonstrated by growing multiple-diode junctions epitaxially on a common substrate that the cutoff frequency could be increased. The CW power capability of his designs is shown in Figure 24.7. η in the figure is conversion efficiency. Peterson and Klemer, [17], show a more common way of combining devices with hybrid couplers. They demonstrated using IMPATT diodes an average power output of about seven times that of Staecker's results at 44 GHz.

24.2.2 Efficiency as a Function of Frequency for Tube and Solid-State RF Sources

Schenk, [19], has shown solid-state device efficiency drops more rapidly with increasing operating frequency than that of traveling-wave tubes. His curves can be summarized by the following approximations; For transistors:

$$
\begin{aligned}
\eta &\approx 120 F_{GHz}^{-0.77} \quad \% , \qquad 4 < F_{GHz} \leq 20 \\
&\approx 52 F_{GHz}^{-0.2} \quad \% , \qquad 1 \leq F_{GHz} \leq 4
\end{aligned}
\tag{24.3}
$$

For traveling-wave-tube amplifiers:

$$
\eta \approx (41 + 23 \log N_C) F_{GHz}^{-0.0021 F_{GHz}} , 3 \leq F_{GHz} \leq 30
\tag{24.4}
$$

where N_C is the number of collector stages, $2 \leq N_C \leq 5$. The above approximation, although specific to a particular traveling-wave tube (TWT) design, generally indicates the effeciency obtainable with other TWTs. Although Weil, [20], does not give efficiency values as a function of frequency, his values are within the limits of equation 24.4. There are exceptions. At K_a band, Staprans and Symons, [21], show three different TWTs to have efficiencies ranging from 5 to 15% below those values predicted by equation 24.4. In addition, one should pay particular attention to not using equation 24.4 below frequencies of 3 GHz. In particular [21] shows an 80% efficiency for a Gyrocon (a special tube) at a frequency of 0.43 GHz, [21]. There are other exceptions, 80% efficiency using a Peniotron (a special tube), [21], at 95 GHz. Tserng et al. [22], show efficiency values in the K_u through the lower millimeter wave frequencies for GaAs MESFET power transistors to more closely follow the TWT efficiency approximating function.

The efficiency of amplifiers depends on the class of operation. The classes of operation are: A, AB, B, and C. In class A load current flows at all times, in class AB load current flows for an appreciable amount but less than an entire cycle of the amplified signal, in class B the load current is zero when input signal is absent and about one-half of each cycle when input signal is present, in class C load current is zero when the signal is absent and is appreciably less than one half of a signal cycle when it is present. The relationship of amplifier class of operation, efficiency and power delivered to the load is shown in Figure 24.8 for transistor amplifiers.

In the figure the power delivered to the load, P_L, is normalized by $P_S = V_{DC}^2 / 2R_S$, where V_{DC} is the dc supply voltage, and R_S is the device saturation resistance. $\eta_c = P_L / P_{DC}$ is the device collector efficiency. P_L is the power delivered to the load and P_{DC} is the power supplied to the amplifier. We see in the figure that the higher output power values are in the upper right-hand corner of the figure and the higher efficiency values are in the lower left area of the figure. Kushner, [23], gives tables of equations for determining optimum load resistance, maximum supply voltage, dissipated power, efficiency and output power for the various classes of amplifiers.

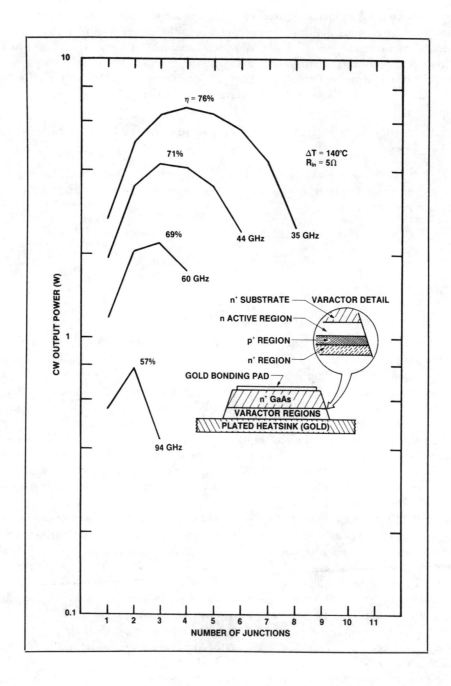

Figure 24.7: CW output power vs. number of junctions for GaAs multiple varactor diodes epitaxially grown on a common substrate. From Staecker [18].

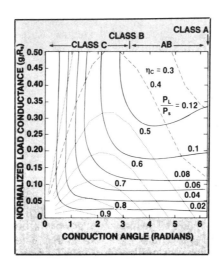

Figure 24.8: Contours of constant power and efficiency on an operating parameter plane. From Adlerstein and Zaitlin [24].

The amplifier power-added efficiency is:

$$\eta_{PAE} = \frac{P_L - P_{IN}}{P_{DC}} = \eta_c \left(1 - \frac{1}{G}\right) \tag{24.5}$$

where P_{IN} is the signal input power and G is the amplifer gain.

The power dissipated by the device is, [25]:

$$P_{dis} \approx P_L(\eta_{PAE}^{-1} - 1) \tag{24.6}$$

for $G \geq 10$.

Comparing central tube with solid-state active aperture

Although vacuum tube power exceeds by far that obtainable from a solid-state rf power source, we must consider the power at the aperture, [25]. For solid-state sources that are distributed with all the antenna elements:

$$P_{avg} = \frac{k_s A_r N_d df L_a P_{cw}(f)}{\lambda^2} \tag{24.7}$$

where

k_s is the number of RF power modules per unit square wavelength; normally ranging from 3 to 4, depending on element spacing.

A_r is the antenna area

N_d is the number of solid-state devices in the RF power source for each module; nominally 2–8.

df   is the duty factor.

L_a is the aperture amplitude distribution loss if the output power of the modules is tapered for low sidelobes.

$P_{cw}(\mathrm{f})$ is the CW power capability of a solid-state device.

For vacuum tube RF power sources:

$$P_{avg} = N_d df L_d P_k(f) \tag{24.8}$$

where N_d is the number of combined tubes, L_d is the loss factor between the transmitter and the aperture, and $P_k(f)$ is the peak-power capability of the tube.

For comparison purposes, one should add to L_d to account for a possibly higher system noise figure for the central power source relative to the active aperture system. In addition one should add to L_a to account for any additional eclipsing loss caused by the normally higher duty factor used in solid-state power systems.

24.2.3   Growth Rate of Average Power Capability

The data in Figure 24.9 show that for the 1 to 10 GHz frequency region tube average power capability is increasing at about 1 dB/year and that solid-state power is increasing at about 0.7 dB/yr at 1 GHz and about 0.4 dB/year at 10 GHz. At a frequency of 100 GHz tube capability is increasing at about 1 2/3 dB/year and solid state at about 1/8 dB/year.

24.2.4   Weight Factor Comparisons

For tubes the weight factor decreases with increasing average power whereas it increases with solid state devices. Woods, [27], shows for microwave tubes the following:

$$w_f \approx 25 P_{avg}^{-0.31} g/W \text{ , for pulsed coupled- cavity TWTs,}$$
$$w_f \approx 38 P_{avg}^{-0.31} \text{ , for cw ECM TWTs} \tag{24.9}$$

The weight factors for millimeter-wave tubes as shown in Table 24.1 are about 11 times that of the appropriate function in equation 24.9 for microwave tubes.

For solid-state average RF power the weight factor is about 2 g/W, varying linearly with P_{avg}, [25]. Equation 2.4 of Chapter 2 gives w_f for T/R modules.

24.3   Spurious Noise, and Phase and Frequency Stability

24.3.1   Spurious Noise

Weil, [20], indicates the in-band spurious noise of linear beam tubes (klystrons and TWTs) is on the order of -90 dBc. For wideband TWTs, Barker, [29] shows values of -83 dBc midband for a tube capable of operation in the frequency range of 2.5 to 7.5 GHz or another tube in the frequency range of 6 to 18 GHz. The spurious noise was -93 dBc when operating the tubes at the edges of the frequency limits. Weil shows the spurious levels of high-gain crossed-field tubes (CFA) to be the order of -70 dBc or -55 dBc for the conventional CFA.

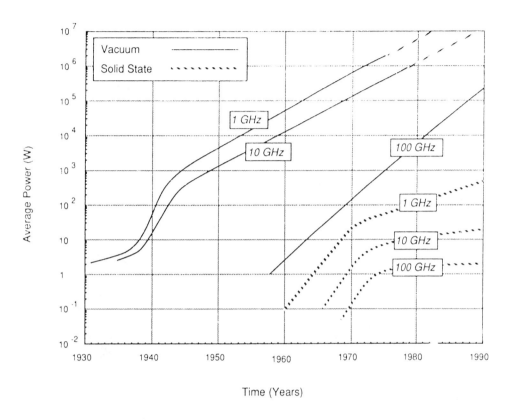

Figure 24.9: Historical growth of solid-state and tube average power. From Sleger et al. [26].

24.3.2 Phase and Amplitude Stability

Short-term stability important for good clutter rejection can be limited by the transmitter phase or frequency stability. Phase and amplitude stability of tubes is rated by pushing factors; the amount of change of phase or amplitude of the amplified signal by changes in voltage applied to the tube electrodes. Millman et al. [28] give example pushing factors for a high-stability TWT, Table 24.2.

Weil, [20] in his Table 4.2, gives representative values of phase change for normalized supply current or voltage changes for various tube types and modulators.

Table 24.2: High-stability TWT amplifier pushing factors, after Millman et al. [28].

Electrode	Phase pushing factor (deg./volt)	Amplitude pushing factor (dB/volt)
Cathode	0.5	0.01
Focus	1	0.02
Collector	0.01	0.0004

Bibliography

[1] Skolnik, M.I., *Introduction to Radar Systems*, McGraw-Hill, New York, 1980.

[2] Borkowski, M.T., "Solid-state transmitters," Ch.5, *Radar Handbook*, M.I. Skolnik, ed., McGraw-Hill, New York, 1990.

[3] Staprans, "Linear Beam Tubes," Ch. 22, *Radar Technology*, E. Brookner, ed., Artech House, Norwood, MA, 1980.

[4] Smith, "Types of crossed-field microwave tubes and their applications," Ch. 23, *Radar Technology*, E. Brookner, ed., Artech House, Norwood, MA, 1980.

[5] Mols, G., C. Corson, K. Lee, "The solid state transmitter for the AN/SPS-40 radar system," IEEE Int. Radar Conf., Arlington, VA, 1985, pp. 76–81.

[6] Ward, H.R., "The RAMP PSR, a solid-state surveillance radar," IEEE Int. Radar Conf., London, 1987, pp. 150–154.

[7] Merrill, P.R., "A 20 kW solid-state L- band transmitter for the RAMP PSR radar," Microwave Journal, March 1988, pp. 165–173.

[8] de Ledinghen, N., L.Wonneberger, "Fully solid-state radar for air traffic control," IEEE Int. Radar Conf., London, 1987, pp. 145–149.

[9] Francoeur, A.R., "Naval Space Surveillance System (NAVSPASUR) solid-state transmitter modernization," IEEE Nat. Radar Conf., Dallas TX 1989, pp. 147–152.

[10] Bhanji, A.M., D.Hoppe, R Cormier, "High power K_a-band transmitter for planetary radar and space craft uplink," IEEE Int. Radar Conf., Arlington, VA, 1085, pp 61–69,

[11] Wallington, J.R., P.J. Chrystie, "A solid-state transmitter with adaptive beamforming," IEEE Int. Radar Conf., Arlington, VA, 1985, pp. 70–75.

[12] Smith, B., "Vacuum electronics in tomorrow's strategic army," Microwave Journal, November 1991, pp. 60–69.

[13] Hansen, J.W., "US TWTs from 1 to 100 GHz," Microwave Journal, September 1989, State of the Art Issue, pp. 179–193.

[14] Belan, M., D. Henry, B. Smith, "High-peak power TWT with very wide bandwidth," Military Microwaves Conf., Microwave Tubes Special Session, London, July 6, 1988, pp. 22–27.

[15] Bierman, H., "mm-Wave devices and subsystems meet military/space demands," Microwave Journal, July 1989, pp. 26–39.

[16] Komiak, J.J., "Design and performance of an octave band 11-W power amplifier MMIC," IEEE Trans. MTT-38, no. 12, December 1990, pp. 2001–2006.

[17] Peterson, D.F., D.P. Klemer, "Multiwatt IMPATT power amplification for EHF applications," Microwave Journal, April 1989, pp. 107–122.

[18] Staecker, P. "mm-Wave transmitters using power frequency multipliers," Microwave Journal, February 1988, pp.175–181.

[19] Schenk, C., "TWT's for new airborne applications," Military Microwaves Conf., Special Session Microwave Tubes, London, July 6, 1988, pp. 17–21.

[20] Weil, T.A., "Transmitters," Ch. 4 *Radar Handbook*, M.I. Skolnik, ed., McGraw-Hill, New York, 1990.

[21] Staprans, A., R.S.Symons, "The 1990 Microwave Power Tube Conference," Microwave Journal, December 1990, pp. 26–39.

[22] Tserng, H.Q., et al. "Millimeter-wave power transistors and circuits," Microwave Journal, April 1989, pp. 125–135.

[23] Kushner, L.J., "Output performance of idealized microwave power amplifiers," Microwave Journal, October 1989, pp. 103–116.

[24] Adlerstein, M.G., M.P.Zaitlin, "Cutoff operation on heterojunction bipolar transistors," Microwave Journal, September 1991, pp. 114–125.

[25] Morchin, W.C., *Airborne Early Warning Radar*, Artech House, Norwood, MA, 1990.

[26] Sleger, K.J., R. Abrams, R. Parker, "Trends in solid-state microwave and millimeter-wave technology," Applied Microwave, Winter 1990, pp. 24–38.

[27] Woods, R.L., "Microwave power source overview," Military Microwaves Conf., London, June 1986, pp. 349–353.

[28] Millman, J.T., J. Saia, C. Hayse, "A high stability TWTA for ground surveillance applications," IEEE Nat. Radar Conf., Atlanta, GA, March 13–14, 1984, pp. 110–114.

[29] Barker, D., "Ultra broadband TWTs for electronic counter measures," Military Microwaves Conf., Special Session Microwave Tubes, London, July 6, 1988, pp. 7–16.

Notes

Notes

Index

MDV 193

Electronic Intelligence: The Analysis of Radar Signals, Richard G. Wiley

Electronic Intelligence: The Interception of Radar Signals, Richard Wiley

Engineer's Refractive Effects Prediction Systems–EREPS, Herbert Hitney

EREPS: Engineer's Refractive Effects Prediction System Software and User's Manual, developed by NOSC

High Resolution Radar, Donald R. Wehner

High Resolution Radar Cross-Section Imaging, Dean Mensa

Interference Suppression Techniques for Microwave Antennas and Transmitters, Ernest R. Freeman

Introduction to Electronic Defense Systems, Filippo Neri

Introduction to Electronic Warfare, D. Curtis Schleher

Introduction to Sensor Systems, S.A. Hovanessian

IONOPROP: Ionospheric Propagation Assessment Software and Documentation, Herbert Hitney

Kalman-Bucy Filters, Karl Brammer

Laser Radar Systems, Albert V. Jelalian

Lidar in Turbulent Atmosphere, V.A. Banakh and V.L. Mironov

Logarithmic Amplification, Richard Smith Hughes

Machine Cryptography and Modern Cryptanalysis, Cither A. Deavours and Louis Kruh

Mathematical Techniques in Multisensor Data Fusion, David L. Hall

Millimeter-Wave Radar Clutter, Nicholas Currie, Robert Hayes, Robert Trebits

Modern Radar System Analysis, David K. Barton

Modern Radar System Analysis Software and User's Manual, David K. Barton and William F. Barton

Monopulse Radar, A.I. Leonov and K.I. Fomichev

MTI and Pulsed Doppler Radar, D. Curtis Schleher

Multifunction Array Radar Design, Dale R. Billetter

Multisensor Data Fusion, Edward L. Waltz and James Llinas

Multiple-Target Tracking with Radar Applications, Samuel S. Blackman

Multitarget-Multisensor Tracking: Advanced Applications, Volume I, Yaakov Bar-Shalom, ed.

Multitarget-Multisensor Tracking: Advanced Applications, Volume II, Yaakov Bar-Shalom, ed.

Over-The-Horizon Radar, A.A. Kolosov, et al.

Surface-Based Air Defense System Analysis, Robert Macfadzean

Surface-Based Air Defense System Analysis Software and User's Guide, Robert Macfadzean and James M. Johnson

Statistical Theory of Extended Radar Targets, R.V. Ostrovityanov and F.A. Basalov

Synthetic-Aperture Radar and Electronic Warfare, Walter W. Goj

VCCALC: Vertical Coverage Plotting Software and User's Manual, John E. Fielding and Gary D. Reynolds